umap

Modules and Monographs in
Undergraduate Mathematics
and its Applications Project

ELEMENTS OF THE THEORY OF GENERALIZED INVERSES FOR MATRICES

Randall E. Cline
Mathematics Department
University of Tennessee
Knoxville, Tennessee 37916

The Project acknowledges Robert M. Thrall,
Chairman of the UMAP Monograph Editorial
Board, for his help in the development and
review of this monograph.

Modules and Monographs in Undergraduate Mathematics
and its Applications Project

The goal of UMAP is to develop, through a community
of users and developers, a system of instructional modules
and monographs in undergraduate mathematics which may
be used to supplement existing courses and from which
complete courses may eventually be built.

The Project is guided by a National Steering Committee
of mathematicians, scientists, and educators. UMAP is funded
by a grant from the National Science Foundation to Education
Development Center, Inc., a publicly supported, nonprofit
corporation engaged in educational research in the U.S. and
abroad.

The Project acknowledges the help of the Monograph
Editorial Board in the development and review of this
monograph. Members of the Monograph Editorial Board
include: Robert M. Thrall, Chairman, of Rice University;
Clayton Aucoin, Clemson University; James C. Frauenthal,
SUNY at Stony Brook; Helen Marcus-Roberts, Montclair
State College; Ben Noble, University of Wisconsin; Paul C.
Rosenbloom, Columbia University. Ex-officio members:
Michael Anbar, SUNY at Buffalo; G. Robert Boynton,
University of Iowa; Kenneth R. Rebman, California State
University; Carroll O. Wilde, Naval Postgraduate School;
Douglas A. Zahn, Florida State University.

The Project wishes to thank Thomas N.E. Greville of
the University of Wisconsin at Madison for his review of
this manuscript and Edwina R. Michener, UMAP Editorial
Consultant, for her assistance.

Project administrative staff: Ross L. Finney, Director;
Solomon Garfunkel, Associate Director/Consortium Coordinator;
Felicia DeMay, Associate Director for Administration; Barbara
Kelczewski, Coordinator for Materials Production.

ISBN-13: 978-0-8176-3013-3   e-ISBN-13: 978-1-4684-6717-8
DOI: 10.1007/978-1-4684-6717-8

TABLE OF CONTENTS

Preface

The purpose of this monograph is to provide a concise
introduction to the theory of generalized inverses of
matrices that is accessible to undergraduate mathematics
majors. Although results from this active area of research
have appeared in a number of excellent graduate level textbooks since 1971, material for use at the undergraduate
level remains fragmented. The basic ideas are so fundamental,
however, that they can be used to unify various topics that
an undergraduate has seen but perhaps not related.

Material in this monograph was first assembled by the
author as lecture notes for the senior seminar in mathematics
at the University of Tennessee. In this seminar one meeting
per week was for a lecture on the subject matter, and another
meeting was to permit students to present solutions to
exercises. Two major problems were encountered the first
quarter the seminar was given. These were that some of the
students had had only the required one-quarter course in
matrix theory and were not sufficiently familiar with
eigenvalues, eigenvectors and related concepts, and that many

of the exercises required fortitude. At the suggestion of
the UMAP Editor, the approach in the present monograph is
(1) to develop the material in terms of full rank factorizations and to relegate all discussions using eigenvalues and
eigenvectors to exercises, and (2) to include an appendix of
hints for exercises. In addition, it was suggested that the
order of presentation be modified to provide some motivation
for considering generalized inverses before developing the
algebraic theory. This has been accomplished by introducing
the Moore-Penrose inverse of a matrix and immediately
exploring its use in characterizing particular solutions to
systems of equations before establishing many of its algebraic properties.

To prepare a monograph of limited length for use at the
undergraduate level precludes giving extensive references to
original sources. Most of the material can be found in
texts such as Ben-Israel and Greville [2] or Rao and Mitra
[11].

Every career is always influenced by colleagues. The
author wishes to express his appreciation particularly to
T.N.E. Greville, L.D. Pyle and R.M. Thrall for continuing
encouragement and availability for consultation.

Randall E. Cline
Knoxville, Tennessee
September 1978

1

Introduction

1.1 Preliminary Remarks

The material in this monograph requires a knowledge of
basic matrix theory available in excellent textbooks such as
Halmos [7], Noble [9] or Strang [12]. Fundamental definitions
and concepts are used without the detailed discussion which
would be included in a self-contained work. Therefore, it may
be helpful to have a standard linear algebra textbook for
reference if needed.

Many examples and exercises are included to illustrate
and complement the topics discussed in the text. It is recom-
mended that every exercise be attempted. Although perhaps
not always successful, the challenge of distinguishing among
what can be assumed, what is known and what must be shown is
an integral part of the development of the nebulous concept
called mathematical maturity.

1.2 Matrix Notation and Terminology

Throughout subsequent sections capital Latin letters denote matrices and small Latin letters denote column vectors. Unless otherwise stated, all matrices (and thus vectors—being matrices having a single column) are assumed to have complex numbers as elements. Also, sizes of matrices are assumed to be arbitrary, subject to conformability in sums and products. For example, writing A+B tacitly assumes A and B have the same size, whereas AB implies that A is m by n and B is n by p for some m, n and p. (Note, however, that even with AB defined, BA is defined if and only if m = p.) The special symbols I and O are used to denote the n by n identity matrix and the m by n null matrix, respectively, with sizes determined by the context when no subscripts are used. If it is important to emphasize size we will write I_n or O_{mn}.

For any $A = (a_{ij})$, the conjugate transpose of A is written as A^H. Thus $A^H = (\bar{a}_{ji})$, where $\bar{a}_{ji}$ denotes the conjugate of the complex scalar a_{ji}, and if x is a column vector with components $x_1,\ldots,x_n$, then x^H is the row vector

$$x^H = (\bar{x}_1,\ldots\bar{x}_n).$$

Consequently, for a real matrix (vector) the superscript "H" denotes transpose.

Given vectors x and y, we write the inner product of x and y as

$$(y,x) = x^H y = \sum_{i=1}^{n} \bar{x}_i y_i.$$

Since only Euclidean norms will be considered, we write $||x||$ without a subscript to mean

$$||x|| = +\sqrt{(x,x)} = +\sqrt{\sum_{i=1}^{n} |x_i|^2}.$$

To conclude this section it is noted that there are certain concepts in the previously cited textbooks which are either used implicitly or discussed in a manner that does not emphasize their importance for present purposes. Although sometimes slightly redundant, the decision to include such topics was based upon the desire to stress fundamental understanding.

Exercises

1.1　Let x be any m-tuple and y be any n-tuple.

　　a.　Form xy^H and yx^H.

　　b.　Suppose m = n and that neither x nor y is the zero vector. Prove that xy^H is Hermitian if and only if $y = \alpha x$ for some real scalar α.

1.2　Let A be any m by n matrix with rows $w_1^H,\ldots,w_m^H$ and columns $x_1,\ldots,x_n$ and let B be any n by p matrix with rows $y_1^H,\ldots,y_n^H$ and columns $z_1,\ldots,z_p$.

　　a.　Prove that the product AB can be written as

$$AB = \begin{bmatrix} (z_1,w_1) & \cdots & (z_p,w_1) \\ \cdot & & \cdot \\ \cdot & & \cdot \\ \cdot & & \cdot \\ (z_1,w_m) & \cdots & (z_p,w_m) \end{bmatrix}$$

　　　　and also as

$$AB = \sum_{i=1}^{n} x_i y_i^H.$$

　　b.　Prove that A = 0 if and only if either $A^H A = 0$ or $AA^H = 0$.

　　c.　Show that $BA^H A = CA^H A$ for any matrices A,B and C implies $BA^H = CA^H$.

*1.3　Let A be any normal matrix with eigenvalues $\lambda_1,\ldots,\lambda_n$ and orthonormal eigenvectors $x_1,\ldots,x_n$.

---

*Exercises or portions of exercises designated by an asterisk assume a knowledge of eigenvalues and eigenvectors.

a. Show that A can be written as

$$A = \sum_{i=1}^{n} \lambda_i x_i x_i^H .$$

b. If $E_i = x_i x_i^H$, i=1,...,n, show that E_i is Hermitian and idempotent, and that $E_i E_j = E_j E_i = 0$ for all $i \neq j$.

c. Use the expression for A in 1.3a and the result of 1.3b to conclude that A is Hermitian if and only if all eigenvalues λ_i are real.

1.3 A Rationale for Generalized Inverses

Given a square matrix, A, the existence of a matrix, X, such that AX = I is but one of many equivalent necessary and sufficient conditions that A is nonsingular. (See Exercise 1.4.) In this case $X = A^{-1}$ is the unique two-sided inverse of A, and $x = A^{-1}b$ is the unique solution of the linear algebraic system of equations Ax = b for every right-hand side b. Loosely speaking, the theory of generalized inverses of matrices is concerned with extending the concept of an inverse of a square nonsingular matrix to singular matrices and, more generally, to rectangular matrices by considering various sets of equations which A and X may be required to satisfy. For this purpose we will use combinations of the following five *fundamental equations*:

(1.1) $A^k XA = A^k$, for some positive integer k,

(1.2) $XAX = X$,

(1.3) $(AX)^H = AX$,

(1.4) $(XA)^H = XA$,

(1.5) $AX = XA$.

(It should be noted that (1.1) with k > 1 and (1.5) implicitly assume A and X are square matrices, whereas (1.1) with k = 1, (1.2), (1.3), and (1.4) require only that X has the size of A^H. Also, observe that all of the equations clearly hold when A

is square and nonsingular, and $X = A^{-1}$.) Given A and subsets
of Equations (1.1)-(1.5), it is logical to ask whether a
solution X exists, is it unique, how can it be constructed
and what properties does it have? These are the basic ques-
tions to be explored in subsequent chapters.

In Chapter 2 we establish the existence and uniqueness
of a particular generalized inverse of any matrix A (to be
called the Moore-Penrose inverse of A), and show how this
inverse can be used to characterize the minimal norm or least
squares solutions to systems of equations Ax = b when A has
full row rank or full column rank. This inverse is then
further explored in Chapter 3 where many of its properties
are derived and certain applications discussed. In Chapter 4
we consider another unique generalized inverse of square
matrices A (called the Drazin inverse of A), and relate this
inverse to Moore-Penrose inverses. The concluding chapter is
to provide a brief introduction to the theory of generalized
inverses that are not unique.

Exercises

1.4   For any A, let N(A) denote the null space of A, that is,

$$N(A) = \{z \mid Az = 0\}.$$

    a.   If A is an n by n matrix, show that the following conditions
are equivalent:

    (i)   A is nonsingular,

    (ii)   N(A) contains only the null vector,

    (iii)   Rank (A) = n,

    (iv)   A has a right inverse,

    (v)   Ax = b has a unique solution for every right-hand
side b.

    b.   What other equivalent statements can be added to this list?

1.5   Let

$$A_4 = \begin{bmatrix} 1 & 2 & 1 & 1 \\ 1 & 1 & 2 & 1 \\ 1 & 1 & 1 & 2 \\ 1 & 1 & 1 & 1 \end{bmatrix}, \quad X = \begin{bmatrix} -1 & -1 & -1 & 4 \\ 1 & 0 & 0 & -1 \\ 0 & 1 & 0 & -1 \\ 0 & 0 & 1 & -1 \end{bmatrix}.$$

a. Show that $X = A_4^{-1}$.

b. If A_5 is the five by five matrix obtained by extending A_4 in the obvious manner, that is, $A_5 = (a_{ij})$ where

$$a_{ij} = \begin{cases} 2, & \text{if } i = j-1, \\ 1, & \text{otherwise}, \end{cases}$$

form A_5^{-1}. More generally, given A_n of this form for any $n \geq 2$, what is A_n^{-1}?

c. Prove that A_n is unimodular for all $n \geq 2$, that is, $|\det A_n| = 1$.

d. Show that any system of equations $A_n x = b$ with b integral has an integral solution x.

1.6 Let x be any vector with $||x|| = 1$ and let k be any real number.

a. Show that $A = I + kxx^H$ is nonsingular for all $k \neq -1$.

b. Given the forty by forty matrix $A = (a_{ij})$ with

$$a_{ij} = \begin{cases} 7, & \text{if } i = j, \\ 1, & \text{otherwise}, \end{cases}$$

construct A^{-1}.

c. Show that A in 1.6a is an involution when $k = -2$.

d. Show that A is idempotent when $k = -1$.

*e. Show that A has one eigenvalue equal to $1+k$ and all other eigenvalues equal to unity. (Hint: Consider x and any vector y orthogonal to x.)

*f. Construct an orthonormal set of eigenvectors for A in 1.6b.

1.7 Given the following pairs of matrices, show that A and X satisfy (1.1) with $k = 1$, (1.2), (1.3), and (1.4).

a.
$$A = \begin{bmatrix} 1 & 0 & 2 \\ -1 & 0 & -2 \end{bmatrix}, \quad X = 1/10 \begin{bmatrix} 1 & -1 \\ 0 & 0 \\ 2 & -2 \end{bmatrix};$$

b.

$$A = \begin{bmatrix} 1 & 2 \\ 2 & -2 \\ -1 & 1 \\ 0 & 1 \end{bmatrix}, \quad X = 1/51 \begin{bmatrix} 16 & 14 & -7 & 3 \\ 15 & -6 & 3 & 6 \end{bmatrix};$$

c.

$$A = \begin{bmatrix} 1 & 2 \\ 3 & 6 \end{bmatrix}, \quad X = 1/50 \begin{bmatrix} 1 & 3 \\ 2 & 6 \end{bmatrix}.$$

1.8   Show that the matrices

$$A = \begin{bmatrix} -4 & -5 & -6 & -4 \\ 1 & 2 & 1 & 1 \\ 2 & 2 & 4 & 2 \\ 1 & 1 & 1 & 1 \end{bmatrix}, \quad X = 1/2 \begin{bmatrix} -3 & -5 & -4 & -3 \\ 2 & 4 & 2 & 2 \\ 1 & 1 & 2 & 1 \\ 0 & 0 & 0 & 0 \end{bmatrix}.$$

satisfy (1.1) with k = 2, (1.2) and (1.5).

2

Systems of Equations and the Moore-Penrose Inverse of a Matrix

2.1   Zero, One or Many Solutions of Ax = b

Given a linear algebraic system of m equations in n
unknowns written as Ax = b, a standard method to determine the
number of solutions is to first reduce the augmented matrix
[A,b] to row echelon form. The number of solutions is then
characterized by relations among the number of unknowns,
rank (A) and rank ([A,b]). In particular, Ax = b is a consistent system of equations, that is, there exists at least one
solution, if and only if rank (A) = rank ([A,b]). Moreover,
a consistent system of equations Ax = b has a unique solution
if and only if rank (A) = n. On the other hand, Ax = b has no
exact solution when rank (A) < rank ([A,b]). It is the
purpose of this chapter to show how the Moore-Penrose inverse
of A can be used to distinguish among these three cases and
to provide alternative forms of representations which are
frequently employed in each case.

For any matrix, A, let CS(A) denote the column space of
A, that is,

$$CS(A) = \{y \,|\, y = Ax, \text{ for some vector } x\},$$

Then $Ax = b$ a consistent system of equations implies $b \epsilon CS(A)$ and conversely (which is simply another way of saying that A and $[A,b]$ have the same rank). Now by definition,

$$\text{rank } (A) = \text{dimension } (CS(A)),$$

and if

$$N(A) = \{z \,|\, Az = 0\}$$

denotes the null space of A (cf. Exercise 1.4), then we have the well-known relation that

$$\text{rank } (A) + \text{dimension } (N(A)) = n.$$

Given a consistent system of equations $Ax = b$ with A m by n of rank r, it follows, therefore, that if $r = n$, then A has full column rank and $x = A_L b$ is the unique solution, where A_L is any left inverse of A. The problem in this case is thus to construct $A_L b$.

However, when $r < n$, so that $N(A)$ consists of more than only the zero vector, then for any solution, x_1, of $Ax = b$, any vector $z \epsilon N(A)$ and any scalar $\alpha, x_2 = x_1 + \alpha z$ is also a solution of $Ax = b$. Conversely, if x_1 and x_2 are any pair of solutions of $Ax = b$, and if $z = x_1 - x_2$, then $Az = Ax_1 - Ax_2 = b - b = 0$ so that $z \epsilon N(A)$. Hence all solutions to $Ax = b$ in this case can be written as

$$(2.1) \qquad x = x_1 + \sum_{i=1}^{n-r} \alpha_i z_i,$$

where x_1 is any particular solution, $z_1, \ldots, z_{n-r}$ are any set of vectors which form a basis of $N(A)$ and $\alpha_1, \ldots, \alpha_{n-r}$ are arbitrary scalars.

Often the problem now is simply to characterize all solutions. More frequently, it is to determine those solutions which satisfy one or more additional conditions as, for example, in linear programming where we wish to construct a nonnegative solution of $Ax = b$ which also maximizes (c,x) where c is some given vector and A, b and c have real elements.

Given an inconsistent system of equations $Ax = b$, that is, where rank $(A) <$ rank $[A,b]$ so that there is no exact solution, a frequently used procedure is to construct a vector $\hat{x}$, say, which is a "best approximate" solution by some criterion. Perhaps the most generally used criterion is that of least squares in which it is required to determine $\hat{x}$ to minimize $||Ax - b||$ or, equivalently, to minimize $||Ax - b||^2$. In this case, if A has full column rank, then $\hat{x} = (A^H A)^{-1} A^H b$ is the least squares solution (see Exercise 2.7).

Exercises

2.1 Given the following matrices, A_i, and vectors, b_i, determine which of the sets of equations $A_i x = b_i$ have a unique solution, infinitely many solutions or no exact solution, and construct the unique solutions when they exist.

(i) $A_1 = \begin{bmatrix} 1 & 2 & 7 \\ -1 & 1 & 2 \\ 4 & -3 & 1 \end{bmatrix}$, $b_1 = \begin{bmatrix} -8 \\ -4 \\ 6 \end{bmatrix}$;

(ii) $A_2 = \begin{bmatrix} 2 & 1 \\ 1 & 0 \\ 3 & 6 \end{bmatrix}$, $b_2 = \begin{bmatrix} 1 \\ 1 \\ 7 \end{bmatrix}$;

(iii) $A_3 = \begin{bmatrix} 3 & 1 & 0 \\ 0 & 3 & 1 \\ 5 & -2 & 5 \\ 1 & 0 & 3 \end{bmatrix}$, $b_3 = \begin{bmatrix} 6 \\ 2 \\ 8 \\ 2 \end{bmatrix}$;

(iv) $A_4 = \begin{bmatrix} 2 & 1 & 1 \\ 3 & 1 & 0 \\ 1 & 0 & -1 \end{bmatrix}$, $b_4 = \begin{bmatrix} 4 \\ 4 \\ 5 \end{bmatrix}$;

(v) $A_5 = \begin{bmatrix} 2 & 1 & 1 \\ 1 & 0 & -1 \\ 3 & 1 & 0 \end{bmatrix}$, $b_5 = \begin{bmatrix} -4 \\ 2 \\ -5 \end{bmatrix}$;

(vi) $A_6 = \begin{bmatrix} 2 & 3 & 1 & 4 \\ 1 & 2 & 0 & 1 \end{bmatrix}$, $b_6 = \begin{bmatrix} 1/2 \\ 0 \end{bmatrix}$.

2.2 For any partitioned matrix $A = [B,R]$ with B nonsingular, prove that columns of the matrix

$$Z = \begin{bmatrix} -B^{-1}R \\ I \end{bmatrix}$$

form a basis of $N(A)$.

2.3 Construct a basis for $N(A_6)$ in Exercise 2.1.

-11-

2.4　Apply the Gram-Schmidt process to the basis in Exercise 2.3 to construct an orthonormal basis of $N(A_6)$.

2.5　Show that if z_1 and z_2 are any vectors which form an orthonormal basis of $N(A_6)$ and if all solutions of $A_6x = b_6$ are written as

$$x = x_1 + \alpha_1 z_1 + \alpha_2 z_2$$

where

$$x_1^H = \frac{1}{12} [0 \quad -1 \quad 1 \quad 2],$$

then $|\alpha_1|^2 + |\alpha_2|^2 = 1$ for every solution such that $||x||^2 = 25/24$.

2.6　Show that A, $A^H A$ and AA^H have the same rank for every matrix A.

2.7　a.　Given any system of equations $Ax = b$ with A m by n and rank $(A) = n$, show by use of calculus that the least squares solution, $\hat{x}$, has the form $\hat{x} = (A^H A)^{-1} A^H b$. Suppose $m = n$?

　　　b.　Construct the least squares solution of $Ax = b$ if

$$A = \begin{bmatrix} 1 & 2 \\ 1 & 0 \\ 1 & 1 \end{bmatrix}, \quad b = \begin{bmatrix} 3 \\ 1 \\ -1 \end{bmatrix}.$$

2.2　Full Rank Factorizations and the Moore-Penrose Inverse

　　Given any matrix A (not necessarily square), it follows at once that if X is any matrix such that A and X satisfy (1.1) with $k = 1$, (1.2), (1.3) and (1.4), then X is unique. For if

(2.2)　　$AXA = A$, $XAX = X$, $(AX)^H = AX$, $(XA)^H = XA$,

and if A and Y also satisfy these equations, then

$$X = XAX = X(AX)^H = XX^H A^H = XX^H (AYA)^H$$
$$= XX^H A^H (AY)^H = XAY = (XA)^H Y = A^H X^H Y$$
$$= (AYA)^H X^H Y = (YA)^H A^H X^H Y = YAXAY = YAY = Y.$$

Now if A has full row rank, then with X any right inverse of A, $AX = I$ is Hermitian and the first two equations in (2.2) hold. Dually, if A has full column rank and X is any left inverse of

A, XA = I is Hermitian and again the first two equations in (2.2) hold. As shown in the following lemma, there is a choice of X in both cases so that all four conditions hold.

LEMMA 1: Let A be any matrix with full row rank or full column rank. If A has full row rank, then $X = A^H(AA^H)^{-1}$ is the unique right inverse of A with XA Hermitian. If A has full column rank, then $X = (A^HA)^{-1}A^H$ is the unique left inverse of A with AX Hermitian.

Proof: If A is any matrix with full row rank, AA^H is non-singular by Exercise 2.6. Now $X = A^H(AA^H)^{-1}$ is a right inverse of A, and

$$(XA)^H = \left[A^H(AA^H)^{-1}A\right]^H = A^H(AA^H)^{-1}A = XA.$$

Thus A and X satisfy the four equations in (2.2), and X is unique.

The dual relationship when A has full column rank follows in an analogous manner with A^HA nonsingular. ∎

It should be noted that $X = A^{-1}$ in (2.2) when A is square and nonsingular, and that both forms for X in Lemma 1 reduce to A^{-1} in this case. More generally, we will see in Theorem 4 that the unique X in (2.2) exists for every matrix A. Such an X is called the *Moore-Penrose inverse* of A and is written A^+. Thus we have from Lemma 1 the special cases:

$$(2.3) \qquad A^+ = \begin{cases} A^H(AA^H)^{-1}, & \text{if A has full row rank,} \\[2mm] (A^HA)^{-1}A^H, & \text{if A has full column rank.} \end{cases}$$

Example 2.1

If $A = \begin{bmatrix} 1 & 0 & 1 \\ 0 & 1 & 1 \end{bmatrix}$ then $(AA^H)^{-1} = \begin{bmatrix} 2 & 1 \\ 1 & 2 \end{bmatrix}^{-1} = \frac{1}{3}\begin{bmatrix} 2 & -1 \\ -1 & 2 \end{bmatrix}$

and so

$$A^+ = \frac{1}{3}\begin{bmatrix} 2 & -1 \\ -1 & 2 \\ 1 & 1 \end{bmatrix}.$$

Example 2.2

   If y is the column vector

$$y = \begin{bmatrix} 1 \\ 2+i \\ 0 \\ -3i \end{bmatrix}$$

then

$$y^+ = \frac{1}{15} \begin{bmatrix} 1 & 2-i & 0 & 3i \end{bmatrix}.$$

Note in general that the Moore-Penrose inverse of any nonzero row or column vector is simply the conjugate transpose of the vector multiplied by the reciprocal of the square of its length.

   Let us next consider the geometry of solving systems of equations in terms of A^+ for the special cases in (2.3). Given any system of equations $Ax = b$ and any matrix, X, such that $AXA = A$, it follows at once that the system is consistent if and only if

(2.4) $AXb = b$.

For if (2.4) holds, then $x = Xb$ is a solution. Conversely, if $Ax = b$ is consistent, multiplying each side on the left by AX gives $Ax = AXAx = AXb$, so that (2.4) follows. Suppose now that $Ax = b$ is a system of m equations in n unknowns where A has full row rank. Then $Ax = b$ is always consistent since (2.4) holds with X any right inverse of A, and we have from (2.1) that all solutions can be written as

(2.5) $x = Xb + Zy$

with Z any matrix with n-m columns which form a basis of $N(A)$ and y an arbitrary vector. Taking X to be the right inverse A^+ in this case gives Theorem 2.

THEOREM 2: For any system of equations $Ax = b$ where A has full row rank, $x = A^+b$ is the unique solution with $||x||^2$ minimal.

-14-

Proof: With $X = A^+$ in (2.5),

$$||x||^2 = (x,x) = (A^+b+Zy, A^+b+Zy)$$
$$= (A^+b, A^+b) + (Zy, Zy) = ||A^+b||^2 + ||Zy||^2$$

since

$$(A^+b, Zy) = (A^H(AA^H)^{-1}b, Zy) = ((AA^H)^{-1}b, AZy) = 0.$$

Thus

$$||x||^2 \geq ||A^+b||^2,$$

where equality holds if and only if $Zy = 0$. ∎

Example 2.3

If A is the matrix in Example 2.1 and $b = \begin{bmatrix} 5 \\ -4 \end{bmatrix}$, then

$$x = A^+b = \frac{1}{3}\begin{bmatrix} 14 \\ -13 \\ 1 \end{bmatrix}$$

is the minimal norm solution of $Ax = b$ with $||x||^2 = \frac{122}{3}$.

It was noted in Section 2.1 that the least squares solution of an inconsistent system of equations $Ax = b$ when A has full column rank is $x = (A^HA)^{-1}A^Hb$. From (2.3) we have, therefore, that $x = A^+b$ is the least squares solution in this case. Although this result can be established by use of calculus (Exercise 2.7), the following derivation in terms of norms is more direct.

THEOREM 3: For any system of equations $Ax = b$ where A has full column rank, $x = A^+b$ is the unique vector with $||b-Ax||^2$ minimal.

Proof: If A is square or if $m > n$ and $Ax = b$ is consistent, then with $A^+ = (A^HA)^{-1}A^H$ a left inverse of A and $AA^+b = b$, the vector $x = A^+b$ is the unique solution with $||b-Ax||^2 = 0$. On the other hand, if $m > n$ and $Ax = b$ is inconsistent,

$$||b-Ax||^2 = ||(I-AA^+)b - A(x-A^+b)||^2$$
$$= ||b-AA^+b||^2 + ||A(x-A^+b)||^2$$

-15-

since $A^H(I-AA^+) = 0$. Hence $||b-Ax||^2 \geq ||b-AA^+b||^2$ where equality holds if and only if $||A(x-A^+b)||^2 = 0$. But A with full column rank implies $||Ay||^2 > 0$ for any vector $y \neq 0$, in particular for $y = x - A^+b$. ∎

Example 2.4

If

$$A = \begin{bmatrix} 2 & -1 \\ 1 & 1 \\ -1 & 0 \end{bmatrix}, \quad b = \begin{bmatrix} 1 \\ 2 \\ 3 \end{bmatrix},$$

then

$$A^+ = \frac{1}{11}\begin{bmatrix} 3 & 3 & -2 \\ -4 & 7 & -1 \end{bmatrix}, \quad AA^+b = \frac{1}{11}\begin{bmatrix} -1 \\ 10 \\ -3 \end{bmatrix} \neq b,$$

and

$$x = A^+b = \frac{1}{11}\begin{bmatrix} 3 \\ 7 \end{bmatrix}$$

is the least squares solution of $Ax = b$ with $||b-Ax||^2 = \frac{144}{11}$ minimal.

Having established A^+ for the special cases in Lemma 1, it remains to establish existence for the general case of an arbitrary matrix A. For this purpose we first require a definition.

DEFINITION 1: Any product EFG with E m by r, F r by r and G r by n is called a *full rank factorization* if each of the matrices E, F and G has rank r.

The importance of Definition 1 is that any nonnull matrix can be expressed in terms of full rank factorizations, and that the Moore-Penrose inverse of such a product is the product of the corresponding inverse in reverse order.

To construct a full rank factorization of a nonnull matrix, let A be any m by n matrix with rank r. Designate columns of A as $a_1,\ldots,a_n$. Then A with rank r implies that there exists at least one set of r columns of A which are linearly independent. Let $J = \{j_1,\ldots,j_r\}$ be any set of indices for which $a_{j_1},\ldots,a_{j_r}$ are linearly independent, and let E be the m by r matrix

$$E = [a_{j_1},\ldots,a_{j_r}].$$

If $r = n$, then $A = E$ is a trivial full rank factorization (with $F = G = I$). Suppose $r < n$. Then for every column $a_j, j \notin J$, there is a column vector y_j, say, such that $a_j = Ey_j$. Now form the r by n matrix, G, with columns $g_1,\ldots,g_n$ as follows: Let

$$g_j = \begin{cases} y_j, & \text{if } j \notin J, \\[2mm] e_i, & \text{if } j = j_i \in J, \end{cases}$$

where $e_i, i=1,\ldots,r$, denote unit vectors. For this matrix G we then have

$$EG = [a_1,\ldots,a_n] = A.$$

Moreover, since the columns $e_1,\ldots,e_r$ of G form a r by r identity matrix, rank $(G) = r$, and with rank $(E) = r$ by construction, $A = EG$ is a full rank factorization (with $F = I$).

That a full rank factorization $A = EFG$ is not unique is apparent by observing that if M and N are any nonsingular matrices, then $A = EM(M^{-1}FN)N^{-1}G$ is also a full rank factorization. The following example illustrates four full rank factorizations of a given matrix, A, where $F = I$ in each case.

Example 2.5

Let
$$A = \begin{bmatrix} 2 & 0 & 4 & 2 & 6 \\ 1 & 1 & 1 & 2 & -1 \\ -1 & 3 & -5 & 2 & -15 \end{bmatrix}.$$

Then

$$A = \begin{bmatrix} 2 & 0 \\ 1 & 1 \\ -1 & 3 \end{bmatrix} \begin{bmatrix} 1 & 0 & 2 & 1 & 3 \\ 0 & 1 & -1 & 1 & -4 \end{bmatrix} = \begin{bmatrix} 2 & 4 \\ 1 & 1 \\ -1 & -5 \end{bmatrix} \begin{bmatrix} 1 & 2 & 0 & 3 & -5 \\ 0 & -1 & 1 & -1 & 4 \end{bmatrix}$$

$$= \begin{bmatrix} 4 & 6 \\ 1 & -1 \\ -5 & -15 \end{bmatrix} \begin{bmatrix} 4/5 & 3/5 & 1 & 7/5 & 0 \\ -1/5 & -2/5 & 0 & -3/5 & 1 \end{bmatrix} = \begin{bmatrix} 0 & 2 \\ 1 & 2 \\ 3 & 2 \end{bmatrix} \begin{bmatrix} -1 & 1 & -3 & 0 & -7 \\ 1 & 0 & 2 & 1 & 3 \end{bmatrix}.$$

Using full rank factorization, the existence of the Moore-Penrose inverse of any matrix follows at once. The following theorem, stated in the form rediscovered by Penrose [10] but originally established by Moore [8], is fundamental to the theory of generalized inverses of matrices.

THEOREM 4: For any matrix, A, the four equations

$$AXA = A, \quad XAX = X, \quad (AX)^H = AX, \quad (XA)^H = XA$$

have a unique solution $X = A^+$. If $A = 0_{mn}$ is the m by n null matrix, $A^+ = 0_{nm}$. If A is not the null matrix, then for any full rank factorization EFG of A, $A^+ = G^+F^{-1}E^+$.

Proof: Uniqueness in every case follows from the remarks after (2.2).

If $A = 0_{mn}$, then $XAX = X$ implies $X = A^+ = 0_{nm}$. If A is not the null matrix, then for any full rank factorization $A = EFG$ it follows by definition that E has full column rank, F is nonsingular and G has full row rank. Thus $E^+ = (E^HE)^{-1}E^H$ and $G^+ = G^H(GG^H)^{-1}$, by (2.3), with E^+ a left inverse of E and G^+ a right inverse of G. Then if $X = G^+F^{-1}E^+$, $XA = G^+G$ and $AX = EE^+$ are Hermitian, by Lemma 1. Moreover, $AXA = A$ and $XAX = X$, so that $X = A^+$. ∎

It should be noted that although the existence of a full rank factorization $A = EG$ has been established for any non-null matrix A, this does not provide a systematic computational procedure for constructing a factorization. Such a procedure will be developed in Exercise 3.3, however, after we have considered the relationship between A^+ and the Moore-Penrose inverse of matrices obtained by permuting rows or columns or both rows and columns of A. Observe, moreover, that if $Ax = b$ is any system of equations with $A = EG$ a full rank factorization, and if $y = Gx$, then $y = E^+b$ is the least squares solution to $Ey = b$, by Theorem 3. Now the system of equations

$Gx = E^+b$ is always consistent and has minimal norm solution $X = G^+E^+b$, by Theorem 2. Consequently we can combine the results of Theorems 2 and 3 by saying that $x = A^+b$ is the least squares solution of $Ax = b$ with $||x||^2$ minimal. Although of mathematical interest (see, for example, Exercises 3.12 and 3.13), most practical applications of least squares require that problems be formulated in such a way that the matrix A has full column rank.

Exercises

2.8 Show that $x_1 = A_6^+b$ in Exercise 2.5.

2.9 Show that if A is any nonsingular matrix, then

$$[A,B]^+ = \begin{bmatrix} A^H(AA^H+BB^H)^{-1} \\ B^H(AA^H+BB^H)^{-1} \end{bmatrix}.$$

2.10 Let u be the column vector with n elements each equal to unity. Show that

$$[I,u]^+ = \frac{1}{n+1} \begin{bmatrix} (n+1)I - uu^H \\ u^H \end{bmatrix}.$$

2.11 a. Given any real numbers $b_1,\ldots,b_n$, show that all solutions to the equations

$$x_i + x_{n+1} = b_i, \quad i = 1,\ldots,n,$$

can be written as

$$x_i = b_i - \frac{1}{n+1}\sum_{i=1}^{n} b_i + \alpha, \quad i = 1,\ldots,n,$$

and

$$x_{n+1} = \frac{1}{n+1}\sum_{i=1}^{m} b_i - \alpha,$$

where α is arbitrary.

b. For what choice of α can we impose the additional condition that

$$\sum_{i=1}^{n} x_i = 0?$$

c. Show that when the condition in 2.11b is imposed, x_{n+1} becomes simply the mean of $b_1, \ldots, b_n$, that is,

$$x_{n+1} = \frac{1}{n} \sum_{i=1}^{n} b_i.$$

d. Show that the problem of solving the equations in 2.11a, subject to the conditions in 2.11b, can be formulated equivalently as a system of equations $Ax = b$ with the $n+1$ by $n+1$ matrix A Hermitian and nonsingular.

2.12 Given any real numbers $b_1, \ldots, b_n$, show that the mean is the least squares solution to the equations

$$x = b_i, \quad i = 1, \ldots, n.$$

2.13 If $Ax = b$ is any system of equations with $A = uv^H$ a matrix of rank one, show that

$$x = \frac{(b,u)}{||u||^2 ||v||^2} v$$

is the least squares solution with minimal norm.

2.14 Let $Ax = b$ be any consistent system of equations and let $z_1, \ldots, z_{n-r}$ be any set of vectors which form an orthonormal basis of $N(A)$, where rank $(A) = r$. Show that if $\overline{x}$ is any solution of $Ax = b$,

$$A^+ b = \overline{x} - \sum_{i=1}^{n-r} \alpha_i z_i$$

with $\alpha_i = (\overline{x}, z_i)$, $i = 1, \ldots, n-r$.

2.15 (Continuation): Let A be any m by n matrix with full row rank, and let Z be any n by $n-m$ matrix whose columns form an orthonormal basis of $N(A)$. Prove that if X is any right inverse of A,

$$A^+ = X - ZZ^H X.$$

2.16 Use the results of Exercises 2.4 and 2.15 to construct A_6^+ starting with the right inverse

$$X = \begin{bmatrix} 2 & -3 \\ -1 & 2 \\ 0 & 0 \\ 0 & 0 \end{bmatrix}.$$

2.3 Some Geometric Illustrations

In this section we illustrate the geometry of Theorems 2 and 3 with some diagrams:

Consider a single equation in three real variables of the form

(2.6) $a_{11}x_1 + a_{12}x_2 + a_{13}x_3 = b_1.$

Then it is well known that all vectors $x^H = [x_1,x_2,x_3]$ which satisfy (2.6) is a plane $P_1(b_1)$, as shown in Figure 1. Now the plane $P_1(0)$ is

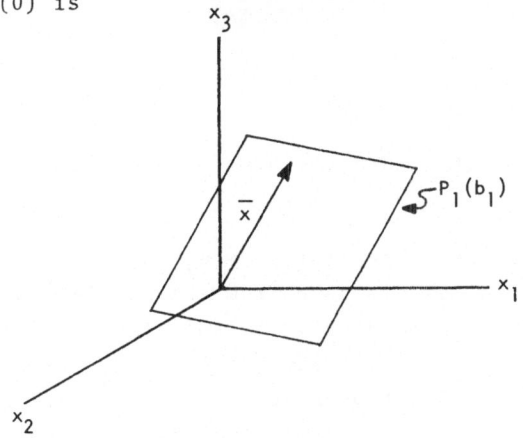

Figure 1. The plane $P_1(b_1)$ and solution $\bar{x}$.

parallel to $P_1(b_1)$, and consists of all solutions $z^H = [z_1,z_2,z_3]$ to the homogeneous equation.

(2.7) $a_{11}z_1 + a_{12}z_2 + a_{13}z_3 = 0.$

Then if $b_1 \neq 0$, all solutions $\bar{\bar{x}}\varepsilon P_1(b_1)$ can be written as $\bar{\bar{x}} = \bar{x} + z$ for some $z\varepsilon P_1(0)$, and conversely, as shown in Figure 2. (Clearly, this is the geometric interpretation of (2.1) for a single equation in three unknowns with two vectors required to span $P_1(0)$.) If we now let $a_1^H = [a_{11},a_{12},a_{13}]$, so that (2.6) can be written as $a_1^H x = b_1$, Theorem 2 implies

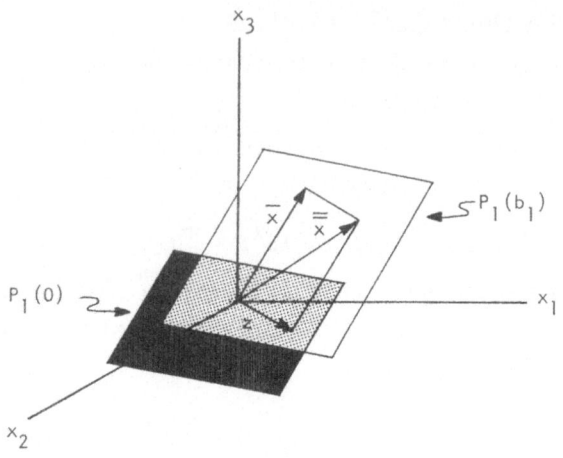

Figure 2. $P_1(b_1)$, $P_1(0)$, $\overline{x}$, z and $\overline{\overline{x}}$.

that the solution of the form $\hat{x} = a_1^{H+}b_1$ is the point on
$P_1(b_1)$ with minimal distance from the origin. Also, since
the vector $\hat{x}$ is perpendicular to the planes $P_1(0), P_1(b_1)$,
$||\hat{x}||$ is the distance between $P_1(0)$ and $P_1(b_1)$. The representation of any solution $\overline{x}$ as $\overline{x} = a_1^{H+}b_1 + \alpha_1 z$, corresponding
to (2.1) in this case, is illustrated in Figure 3.

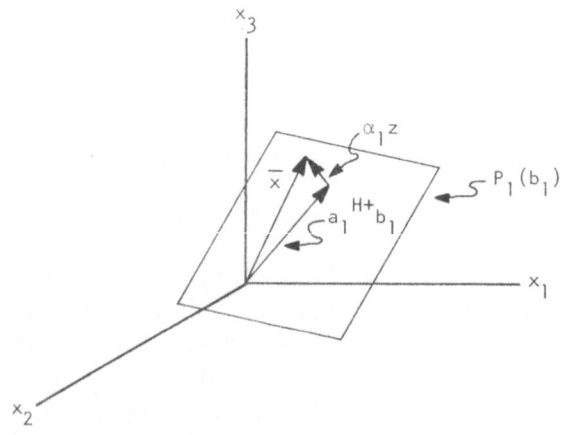

Figure 3. The representation $\overline{x} = a_1^{H+}b_1 + \alpha_1 z$.

Suppose next that we consider (2.6) and a second equation

(2.8) $\qquad a_{21}x_1 + a_{22}x_2 + a_{23}x_3 = b_2$.

Let the plane of solutions of (2.8) be designated as $P_2(b_2)$.
Then it follows that either the planes $P_1(b_1)$ and $P_2(b_2)$ coincide, or they are parallel and distinct, or they intersect in
a straight line. In the first case, when $P_1(b_1)$ and $P_2(b_2)$
coincide, the equation in (2.8) is a multiple of (2.6) and
any point satisfying one equation also satisfies the other.
On the other hand, when $P_1(b_1)$ and $P_2(b_2)$ are parallel and
distinct, there is no exact solution. Finally, when $P_1(b_1)$
and $P_2(b_2)$ intersect in a straight line ℓ_{12}, say, that is,
$\ell_{12} = P_1(b_1) \cap P_2(b_2)$, then any point on ℓ_{12} satisfies both
(2.6) and (2.8). Observe, moreover, that with

$$A = \begin{bmatrix} a_{11} & a_{12} & a_{13} \\ a_{21} & a_{22} & a_{23} \end{bmatrix}, \quad b = \begin{bmatrix} b_1 \\ b_2 \end{bmatrix},$$

the point on ℓ_{12} with minimal distance from the origin is
$\hat{x} = A^+b$. This last case is illustrated in Figure 4, where
ℓ_{12} is a "translation" of the subspace $N(A)$ of the form
$P_1(0) \cap P_2(0)$.

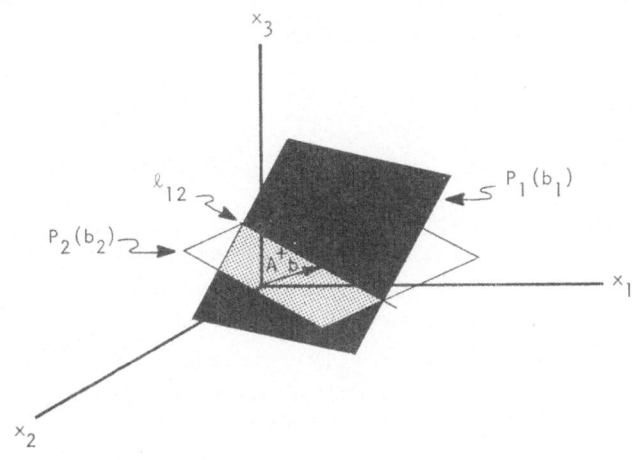

Figure 4. $P_1(b_1)$, $P_2(b_2)$, ℓ_{12} and A^+b.

The extension to three or more equations is now obvious: Given a third equation

(2.9) $a_{31}x_1 + a_{32}x_2 + a_{33}x_3 = b_3$,

let $P_3(b_3)$ be the associated plane of solutions. Then assuming the planes $P_1(b_1)$ and $P_2(b_2)$ do not coincide or are not parallel and distinct, that is, they intersect in the line ℓ_{12} as shown in Figure 4, the existence of a vector $x^H = (x_1, x_2, x_3)$ satisfying (2.6), (2.8) and (2.9) is determined by the conditions that either $P_3(b_3)$ contains the line ℓ_{12}, or $P_3(b_3)$ and ℓ_{12} are parallel and distinct, or $P_3(b_3)$ and ℓ_{12} intersect in a single point. (The reader is urged to construct figures to illustrate these cases. An illustration of three different planes containing the same line may also be found in Figure 5.) For $m \geq 4$ equations, similar considerations as to the intersections of planes $P_k(b_k)$ and lines $\ell_{ij} = P_i(b_i) \cap P_j(b_j)$ again hold, but diagrams become exceedingly difficult to visualize.

For any system of equations $Ax = b$ let $y = AA^+b$, that is, y is the perpendicular projection of b onto $CS(A)$, the column space of A. Then it follows from (2.4) that $Ax = y$ is always a consistent system of equations, and from Theorem 2 that $A^+y = A^+(AA^+)b = A^+b$ is the minimal norm solution. Moreover, we have from Theorem 3 that if $Ax = b$ is inconsistent, then

$$||Ax-b||^2 = ||y-b||^2 = ||AA^+b-b||^2$$

is minimal. Thus, the minimal norm solution A^+y of $Ax = y$ also minimizes $||y-b||^2$.

Consider an inconsistent system of, say, three equations in two unknowns, $Ax = b$, and suppose rank $(A) = 2$. Let $y = AA^+b$ have components y_1, y_2, y_3, and let a_1^H, a_2^H, a_3^H designate rows of A. Now if $P_i(y_i)$ is the plane of all solutions of $a_i^H x = y_i$, $i = 1,2,3$, then the set of all solutions of $Ax = y$ is the line $\ell_{12} = P_1(y_1) \cap P_2(y_2)$ as shown in Figure 5a, $A^+y = A^+b$ is the point on ℓ_{12} of minimal norm and $||y-b||^2$ is minimal, as shown in Figure 5b.

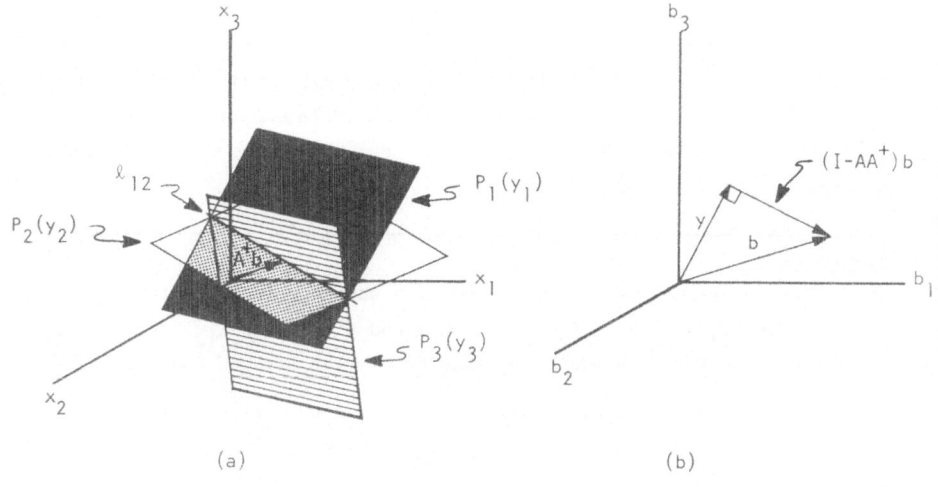

Figure 5. (a) Solutions of $Ax = y$ where $y = AA^+b$.

(b) The vectors b, y and $(I-AA^+)b$.

To conclude this section we remark that since

$$b = AA^+b + (I-AA^+)b$$

is an orthogonal decomposition of any vector b with

$$||b||^2 = ||AA^+b||^2 + ||(I-AA^+)b||^2,$$

then the ratio

$$(2.10) \qquad \phi = \frac{||AA^+b||^2}{||(I-AA^+)b||^2} \geq 0$$

provides a measure of inconsistency of the system $Ax = b$. In particular, $\phi = 0$ implies b is orthogonal to $CS(A)$, whereas large values of ϕ imply that b is nearly contained in $CS(A)$, that is, $||(I-AA^+)b||^2$ is relatively small. (For statistical applications [1] [4] [5], the values $||b||^2, ||AA^+b||^2$ and $||(I-AA^+)b||^2$ are frequently referred to as TSS [Total sum of squares], SSR [Sum of squares due to regression] and SSE [Sum of squares due to error], respectively. Under certain general assumptions, particular multiples of ϕ can be shown to have distributions which can be used in tests of significance.)

-25-

Although the statistical theory of linear regression models is not germane to the present considerations, formation of ϕ in (2.10) can provide insight into the inconsistency of a system of equations $Ax = b$. (See Exercise 2.18.)

Exercises

2.17 Use the techniques of solid analytic geometry to prove that the lines $\ell_{12} = P_1(b_1) \cap P_2(b_2)$, $b_1 \neq 0$ and $b_2 \neq 0$ and $\overline{\ell}_{12} = P_1(0) \cap P_2(0)$ are parallel. In addition, show by similar methods that if

$$P_i(b_i) = \{x \mid a_i^H x = b_i, \; a_i^H = (a_{i1}, a_{i2}, a_{i3})\},$$

then

$$||a_i^{H+} b_i||^2 = \min ||x||^2$$
$$x \epsilon P_i(b_i).$$

2.18 Given any points (x_i, y_i), $i = 0, 1, \ldots, n$, in the (x,y) plane with $x_0, \ldots, x_n$ distinct, it is well known that there is a unique interpolating polynomial $P_n(x)$ of degree $\leq n$ (that is, $P_n(x_i) = y_i$ for all $i = 0, \ldots, n$), and if

$$P_n(x) = \alpha_0 + \alpha_1 x + \ldots + \alpha_n x^n$$

then $\alpha_0, \ldots, \alpha_n$ can be determined by solving the system of equations $A\alpha = y$ where

$$A = \begin{bmatrix} 1 & x_0 & \cdots & x_0^n \\ 1 & x_1 & \cdots & x_1^n \\ \cdot & \cdot & & \cdot \\ \cdot & \cdot & & \cdot \\ \cdot & \cdot & & \cdot \\ 1 & x_n & \cdots & x_n^n \end{bmatrix}, \quad \alpha = \begin{bmatrix} \alpha_0 \\ \alpha_1 \\ \cdot \\ \cdot \\ \cdot \\ \alpha_n \end{bmatrix}, \quad y = \begin{bmatrix} y_0 \\ y_1 \\ \cdot \\ \cdot \\ \cdot \\ y_n \end{bmatrix}.$$

Now any matrix, A, with this form is called a Vandermonde matrix, and it can be shown that

$$\det(A) = \prod_{i<j} (x_i - x_j).$$

Thus, with $x_0,\ldots,x_n$ distinct, A is nonsingular, and if A_k denotes the submatrix consisting of the first k columns of A, $k = 1,2,\ldots,n$, then A_k has full column rank for every k.

For $k \leq n$, the least squares polynomial approximation of degree k to the points (x_i, y_i), $i = 0,\ldots,n$, is defined to be that polynomial

$$P_k(x) = \alpha_0 + \alpha_1 x + \ldots \alpha_k x^k$$

which minimizes

$$\sum_{i=0}^{n} [y_i - P_k(x_i)]^2.$$

a.   Show that the coefficients $\alpha_0,\ldots,\alpha_k$ of the least squares polynomial approximation of degree k are elements of the vector α, where

$$\alpha = A_{k+1}^{+} y.$$

b.   Show that with TSS $= \sum_{i=0}^{n} y_i^2$, then SSR $= ||A_{k+1} A_{k+1}^{+} y||^2$.

c.   Given the data

x_i	-1	0	1	2
y_i	-5	-4	-3	10

construct the best linear, quadratic and cubic least squares approximations. For each case determine SSR and SSE. What conclusions can you draw from the data available?

2.4   Miscellaneous Exercises

2.19   Let A, Z_1 and Z_2 be any matrices.

a.   Prove that a solution, X, to the equations $XAX = X$, $AX = Z_1$ and $XA = Z_2$, if it exists, is unique.

b.   For what choices of Z_1 and Z_2 is X a generalized inverse of A?

2.20   Verify the following steps in the original Penrose proof of the existence of X in (2.2):

a. The equations of $XAX = X$ and $(AX)^H = AX$ are equivalent to the single equations $XX^H A^H = X$. Dually, $AXA = A$ and $(XA)^H = XA$ are equivalent to the single equation $XAA^H = A^H$.

b. If there exists a matrix B satisfying $BA^H AA^H = A^H$, then $X = BA^H$ is a solution of the equations $XX^H A^H = X$ and $XAA^H = A^H$.

c. The matrices $A^H A$, $(A^H A)^2$, $(A^H A)^3,\ldots$, are not all linearly independent, so that there exists scalars $d_1,\ldots,d_k$ not all zero, for which

$$d_1 A^H A + d_2 (A^H A)^2 + \ldots + d_k (A^H A) = 0.$$

(Note that if A has n columns, $k \le n^2+1$. Why?)

d. Let d_s be the first nonzero scalar in the matrix polynomial in 2.20c, and let

$$B = \frac{1}{d_s} \{d_{s+1} I + d_{s+2} A^H A + \ldots + d_k (A^H A)^{k-s-1}\}.$$

Then $B(A^H A)^{s+1} = (A^H A)^s$.

e. The matrix B also satisfies $BA^H AA^H = A^H$.

2.21 Let A and X be any matrices such that $AXA = A$. Show that if $Ax = b$ is a consistent system of equations, then all solutions can be written as

$$x = Xb + (I-XA)y$$

where y is arbitrary. (Note, in particular, that this expression is equivalent to the form for x in (2.5) since columns of $I-XA$ form a basis for $N(A)$. Why?)

2.22 (Continuation): Prove, more generally, that $AWC = B$ is a consistent system of equations if and only if $AA^+ BC^+ C = B$, in which case all solutions can be written as

$$W = A^+ BC^+ + Y - A^+ AYCC^+$$

where Y is arbitrary.

3

More on
Moore-Penrose Inverses

3.1 Basic Properties of A^+

The various properties of A^+ discussed in this section
are fundamental to the theory of Moore-Penrose inverses. In
many cases, proofs simply require verification that the
defining equations in (2.2) are satisfied for A and some
particular matrix X. Having illustrated this proof technique
in a number of cases, we will leave the remaining similar
arguments as exercises.

LEMMA 5: Let A be any m by n matrix. Then

(a) A m by n implies A^+ n by m;

(b) $A = 0_{mn}$ implies $A^+ = 0_{nm}$;

(c) $A^{++} = A$;

(d) $A^{H+} = A^{+H}$;

(e) $A^+ = (A^H A)^+ A^H = A^H (AA^H)^+$;

(f) $(A^H A)^+ = A^+ A^{H+}$;

(g) $(\alpha A)^+ = \alpha^+ A^+$ for any scalar α, where

$$\alpha^+ = \begin{cases} 1/\alpha, & \text{if } \alpha \neq 0, \\ 0, & \text{if } \alpha = 0; \end{cases}$$

(h) If U and V are unitary matrices, $(UAV)^+ = V^H A^+ U^H$;

(i) If $A = \sum_{i=1}^{n} A_i$ where $A_i^H A_j = 0$ whenever $i \neq j$, $A^+ = \sum_{i=1}^{n} A_i^+$;

(j) If A is normal, $A^+ A = AA^+$;

(k) A, A^+, $A^+ A$ and AA^+ all have rank equal to trace $(A^+ A)$.

Proof: Properties (a) and (b) have been noted previously in Section 1.3 and Theorem 4, respectively. The relations in (c) and (d) follow by observing that there is complete duality in the roles of A and X in the defining equations.

To establish the first expression for A^+ in (e), let $X = (A^H A)^+ A^H$. Then $XA = (A^H A)^+ A^H A$ is Hermitian, and also $AX = A(A^H A)^+ A^H$ by use of (d). Moreover, $XAX = X$ and

$$AXA = A(A^H A)^+ A^H A = A^{H+} A^H A (A^H A)^+ A^H A = A^{H+} A^H A = A.$$

The second expression in (e) follows by a similar type of argument, as do the expressions in (g) and (h).

To prove (f) we have $A^{H+} = A(A^H A)^+$ by (d) and (e). Then

$$A^+ A^{H+} = (A^H A)^+ A^H A (A^H A)^+ = (A^H A)^+.$$

To prove (i), observe first that $A_i^H A_j = 0$ implies

$$A_i^+ A_j = A_i^+ A_i^{+H} A_i^H A_j = 0$$

and also $A_j^+ A_i = 0$ since

$$A_j^H A_i = 0.$$

Now we can again show that A and A^+ satisfy the defining equation.

That (j) holds follows by use of (e) to write

$$A^+ A = (A^H A)^+ A^H A = (AA^H)^+ AA^H = A^{H+} A^H = (AA^+)^H = AA^+.$$

To show that A, A^+, A^+A and AA^+ all have the same rank, we can apply the fact that the rank of a product of matrices never exceeds the rank of any factor to the equations $AA^+A = A$ and $A^+AA^+ = A^+$. Then rank (A) = trace (A^+A) holds since rank (E) = trace (E) for any idempotent matrix E [7]. ∎

Observe in Lemma 5(e) that these expressions for A^+ reduce to the expressions in (2.3) whenever A has full row rank or full column rank. Moreover, observe that the relationship $(EG)^+ = G^+E^+$ which holds for full rank factorizations EG, by Theorem 4, also holds for $\Lambda^H A$ where A is any matrix, by Lemma 5(f). The following example shows, however, that the relation $(BA)^+ = A^+B^+$ need not hold for arbitrary matrices A and B.

Example 3.1

Let

$$A = \begin{bmatrix} 1 & 0 \\ 1 & 1 \\ 1 & 1 \end{bmatrix}, \quad B = \begin{bmatrix} 1 & 1 & -1 \\ 0 & 1 & -1 \end{bmatrix}.$$

Then

$$BA = \begin{bmatrix} 1 & 0 \\ 0 & 0 \end{bmatrix} = (BA)^+,$$

since BA is Hermitian and idempotent. Also, we have

$$A^+ = (A^HA)^{-1}A^H = 1/2 \begin{bmatrix} 2 & -2 \\ -2 & 3 \end{bmatrix} \begin{bmatrix} 1 & 1 & 1 \\ 0 & 1 & 1 \end{bmatrix} = 1/2 \begin{bmatrix} 2 & 0 & 0 \\ -2 & 1 & 1 \end{bmatrix}$$

and

$$B^+ = B^H(BB^H)^{-1} = 1/2 \begin{bmatrix} 1 & 0 \\ 1 & 1 \\ -1 & -1 \end{bmatrix} \begin{bmatrix} 2 & -2 \\ -2 & 3 \end{bmatrix} = 1/2 \begin{bmatrix} 2 & -2 \\ 0 & 1 \\ 0 & -1 \end{bmatrix},$$

so that

$$A^+B^+ = \begin{bmatrix} 1 & -1 \\ -1 & 1 \end{bmatrix} \neq (BA)^+.$$

Let A be any m by n matrix with columns $a_1,\ldots,a_n$, and let Q designate the permutation matrix obtained by permuting columns of I_n in any order $\{j_1,\ldots,j_n\}$. Then

$$AQ = [a_{j_1}, \ldots, a_{j_n}].$$

In a similar manner, if $w_1^H, \ldots w_m^H$ designate the rows of A, and if P is the permutation matrix obtained by permuting rows of I_m in any order $\{i_1, \ldots, i_m\}$, then

$$PA = \begin{bmatrix} w_{i_1}^H \\ \cdot \\ \cdot \\ \cdot \\ w_{i_m}^H \end{bmatrix}.$$

Combining these observations it follows, therefore, that if $\tilde{A}$ is any m by n matrix formed by permuting rows of A or columns of A or both rows and also columns of A in any manner, then $\tilde{A}$ = PAQ for some permutation matrices P and Q. Moreover, since P and Q are unitary matrices,

$$\tilde{A}^+ = Q^H A^+ P^H,$$

by Lemma 5(h), and thus

$$A^+ = Q\tilde{A}^+ P.$$

In other words, A^+ can be obtained by permuting rows and/or columns of $\tilde{A}^+$.

Example 3.2

Construct B^+ if

$$B = \begin{bmatrix} 1 & 1 & 0 \\ 1 & 0 & 1 \end{bmatrix}.$$

Since B in this case can be written as

$$B = PAQ = \begin{bmatrix} 0 & 1 \\ 1 & 0 \end{bmatrix} \begin{bmatrix} 1 & 0 & 1 \\ 0 & 1 & 1 \end{bmatrix} \begin{bmatrix} 0 & 0 & 1 \\ 0 & 1 & 0 \\ 1 & 0 & 0 \end{bmatrix}$$

where A is the matrix in Example 2.1, then with P and Q Hermitian

$$B^+ = Q^H A^+ P^H = 1/3 \begin{bmatrix} 0 & 0 & 1 \\ 0 & 1 & 0 \\ 1 & 0 & 0 \end{bmatrix} \begin{bmatrix} 2 & -1 \\ -1 & 2 \\ 1 & 1 \end{bmatrix} \begin{bmatrix} 0 & 1 \\ 1 & 0 \end{bmatrix} = \frac{1}{3} \begin{bmatrix} 1 & 1 \\ 2 & -1 \\ -1 & 2 \end{bmatrix}.$$

(It should be noted that B can be written alternately as

$$B = AQ_1 = \begin{bmatrix} 1 & 0 & 1 \\ 0 & 1 & 1 \end{bmatrix} \begin{bmatrix} 0 & 1 & 0 \\ 0 & 0 & 1 \\ 1 & 0 & 0 \end{bmatrix}$$

and so we have also $B^+ = Q_1^H A^+$.)

Further applications of full rank factorizations and of permuted matrices PAQ in the computation of A^+ will be illustrated in the exercises at the end of this section. We turn now to a somewhat different method for computing A^+. This procedure essentially provides a method for constructing the Moore-Penrose inverse of any matrix with k columns, given that the Moore-Penrose inverse of the submatrix consisting of the first k-1 columns is known.

For any $k \geq 2$, let A_k denote the matrix with k columns, $a_1, \ldots, a_k$. Then A_k can be written in partitioned form as $A_k = [A_{k-1}, a_k]$. Assuming A_{k-1}^+ is known, A_k can be formed using the formulas in Theorem 6.

THEOREM 6: For any matrix $A_k = [A_{k-1}, a_k]$, let

$$c_k = (I - A_{k-1}A_{k-1}^+)a_k$$

and let

$$\gamma_k = a_k^H A_{k-1}^{H+} A_{k-1}^+ a_k.$$

Then

$$(3.1) \qquad A_k^+ = \begin{bmatrix} A_{k-1}^+ - A_{k-1}^+ a_k b_k \\ b_k \end{bmatrix}$$

where

$$b_k = \begin{cases} c_k^+, & \text{if } c_k \neq 0, \\ (1+\gamma_k)^{-1} a_k^H A_{k-1}^{H+} A_{k-1}^+, & \text{if } c_k = 0. \end{cases}$$

Proof: Since c_k is a column vector, the two cases $c_k \neq 0$ and $c_k = 0$ are exhaustive.

Let X designate the right-hand side of (3.1). Then to establish the representation for A_k^+ requires only that we show the defining equations in (2.2) are satisfied by A_k and X for the two forms of b_k.

Forming $A_k X$ and $X A_k$ gives

$$(3.2) \qquad A_k X = A_{k-1} A_{k-1}^+ + c_k b_k,$$

by definition of c_k, and

$$(3.3) \qquad X A_k = \begin{bmatrix} A_{k-1}^+ A_{k-1} - A_{k-1}^+ a_k b_k A_{k-1} & A_{k-1}^+ a_k (I - b_k a_k) \\ b_k A_{k-1} & b_k a_k \end{bmatrix}$$

Continuing, using (3.2) gives

$$(3.4) \qquad A_k X A_k = \begin{bmatrix} A_{k-1} + c_k b_k A_{k-1}, & A_{k-1} A_{k-1}^+ a_k + c_k b_k a_k \end{bmatrix}$$

and

$$(3.5) \qquad X A_k X = \begin{bmatrix} A_{k-1}^+ - A_{k-1}^+ a_k b_k A_{k-1} A_{k-1}^+ - A_{k-1}^+ a_k b_k c_k b_k \\ b_k A_{k-1} A_{k-1}^+ + c_k b_k c_k \end{bmatrix}.$$

since $A_{k-1}^+ c_k = 0$.

Suppose now that $c_k \neq 0$ and $b_k = c_k^+$. Then

$$A_k X = A_{k-1} A_{k-1}^+ + c_k c_k^+$$

in (3.2) is Hermitian. Also, with $c_k^+ c_k = 1$, and since $A_{k-1}^+ c_k = 0$ implies $c_k^H A_{k-1}^{H+} = 0$ so that $c_k^+ A_{k-1} = 0$ and thus $c_k^+ a_k = 1$, then

$$X A_k = \begin{bmatrix} A_{k-1}^+ A_{k-1} & 0 \\ 0 & 1 \end{bmatrix}$$

in (3.3) is Hermitian. Moreover,

-34-

$$A_k X A_k = \left[A_{k-1}, A_{k-1} A_{k-1}^+ a_k + c_k \right] = \left[A_{k-1}, a_k \right] = A_k$$

in (3.4), and

$$XA_k X = \begin{bmatrix} A_k^{-1^+} - A_{k-1}^+ a_k b_k \\ b_k \end{bmatrix} = X$$

in (3.5). Having shown that the defining equations hold, then $X = A_k^+$ in (3.1) when $c_k \neq 0$.

Suppose $c_k = 0$ and $b_k = (1+\gamma_k)^{-1} a_k^H A_{k-1}^{H+} A_{k-1}^+$. Then

$$A_k X = A_{k-1} A_{k-1}^+$$

in (3.2) is Hermitian. In this case, with

$$\gamma_k = a_k^H A_{k-1}^{H+} A_{k-1}^+ a_k$$

a nonnegative real number and

$$b_k a_k = (1+\gamma_k)^{-1} \gamma_k = 1 - (1+\gamma_k)^{-1},$$

we have also

$$XA_k = \begin{bmatrix} A_{k-1}^+ A_{k-1} - (1+\gamma_k)^{-1} A_{k-1}^+ a_k a_k^H A_{k-1}^{H+} & (1+\gamma_k)^{-1} A_{k-1}^+ a_k \\ (1+\gamma_k)^{-1} a_k^H A_{k-1}^{H+} & 1 - (1+\gamma_k)^{-1} \end{bmatrix}$$

in (3.3) Hermitian. Furthermore, with $b_k A_{k-1} A_{k-1}^+ = b_k$ and, since $c_k = 0$ implies $A_{k-1} A_{k-1}^+ a_k = a_k$,

$$A_k X A_k = A_k$$

in (3.4) and $XA_k X = X$ in (3.5). Thus, when $c_k = 0$ it has been shown again that A_k and X satisfy the defining equations for the given form for b_k. ∎

That the formulas in Theorem 6 can be used not only directly to construct A_k^+, assuming A_{k-1}^+ is known, but also recursively to form A^+ for any matrix A is easily seen: Let A be any matrix with n columns $a_1, \ldots, a_n$, and for $k = 1, \ldots, n$,

let A_k designate the submatrix consisting of the first k
columns of A. Now $A_1^+ = a_1^+$ follows directly from Lemma 5(b)
or (e), and if $n \geq 2$, $A_2^+, \ldots, A_n^+ = A^+$ can be formed
sequentially using Theorem 6.

Example 3.3

Let

$$A = \begin{bmatrix} 2 & 1 & 0 \\ 0 & 2 & -4 \\ -1 & 1 & -3 \end{bmatrix}.$$

Then with $A_1 = a_1$,

$$A_1^+ = 1/5 [2 \quad 0 \quad -1],$$

$$A_1^+ a_2 = 1/5$$

and

$$c_2 = a_2 - A_1(A_1^+ a_2) = \begin{bmatrix} 1 \\ 2 \\ 1 \end{bmatrix} - 1/5 \begin{bmatrix} 2 \\ 0 \\ -1 \end{bmatrix} = 1/5 \begin{bmatrix} 3 \\ 10 \\ 6 \end{bmatrix}.$$

Hence

$$b_2 = c_2^+ = 1/29 [3 \quad 10 \quad 6]$$

and so

$$A_2^+ = \begin{bmatrix} 1/5 [2 \quad 0 \quad -1] - 1/145 [3 \quad 10 \quad 6] \\ 1/29 [3 \quad 10 \quad 6] \end{bmatrix}$$

$$= 1/145 \begin{bmatrix} 55 & -10 & -35 \\ 15 & 50 & 30 \end{bmatrix} = 1/29 \begin{bmatrix} 11 & -2 & -7 \\ 3 & 10 & 6 \end{bmatrix}.$$

Continuing,

$$A_2^+ a_3 = 1/29 \begin{bmatrix} 29 \\ -58 \end{bmatrix} = \begin{bmatrix} 1 \\ -2 \end{bmatrix}$$

and

$$c_3 = a_3 - A_2(A_2^+ a_3) = a_3 - a_3 = 0.$$

Thus, with $\gamma_3 = 5$ and

$$a_3{}^H A_2{}^{H+} A^+{}_2 = (A_2{}^+ a_3)^H A_2{}^+ = 1/29 [5 \quad -22 \quad -19],$$

we have

$$b_3 = 1/174 [5 \quad -22 \quad -19]$$

so that

$$A_3{}^+ = \begin{bmatrix} 1/29 \begin{bmatrix} 11 & -2 & -7 \\ 3 & 10 & 6 \end{bmatrix} - 1/174 \begin{bmatrix} 5 & -22 & -19 \\ -10 & 44 & 38 \end{bmatrix} \\ 1/174 [5 \quad -22 \quad -19] \end{bmatrix}$$

$$= 1/174 \begin{bmatrix} 61 & 10 & -23 \\ 28 & 16 & -2 \\ 5 & -22 & -19 \end{bmatrix}.$$

As will be indicated in Exercise 3.7, there is a converse of Theorem 6 which can be used to construct $A_{k-1}{}^+$, given $[A_{k-1}, a_k]^+$. Combining Theorem 6 and its converse thus provides a technique for constructing the Moore-Penrose inverse of a matrix, $\bar{A}$, say, starting from any matrix, A, of the same size with A^+ known. (For practical purposes, however, $\bar{A}$ and A should differ in a small number of columns.

Exercises

3.1 Let A be any matrix with columns $a_1, \ldots, a_n$, and let A^+ have rows $w_1{}^H, \ldots, w_n{}^H$. Prove that if K denotes any subset of the indices $1, \ldots, n$ such that $a_i = 0$, then $w_i{}^H = 0$ for all $i \epsilon K$.

3.2 Let A be any m by n matrix with rank r, $0 < r < \min(m,n)$.

    a.   Prove that there exist permutation matrices, P and Q, such that $\tilde{A} = PAQ$ has the partitioned form

$$\tilde{A} = \begin{bmatrix} W & X \\ Y & Z \end{bmatrix}$$

        with W r by r and nonsingular.

    b.   Show that $Z = YW^{-1}X$.

    c.   Construct $\tilde{A}^+$.

3.3 (Continuation): A matrix [U,V] is called upper trapezoidal if U is upper triangular and nonsingular; a matrix, B, is called lower trapezoidal if B^H is upper trapezoidal.

a. Show that any matrix, $\tilde{A}$, in Exercise 3.2 has a full rank factorization $\tilde{A} = EG$ with E lower trapezoidal and G upper trapezoidal. (Such a factorization $\tilde{A} = EG$ is called a *trapezoidal decomposition* of $\tilde{A}$.)

b. Construct a trapezoidal decomposition of some matrix, $\tilde{A}$, obtained from

$$A = \begin{bmatrix} 1 & 2 & -1 & 0 \\ 2 & 4 & 0 & 2 \\ 3 & 6 & 1 & 4 \end{bmatrix}.$$

Hint: Start with

$$A = \begin{bmatrix} 1 & 0 & 0 \\ 2 & * & 0 \\ 3 & * & * \end{bmatrix} \begin{bmatrix} 1 & 2 & -1 & 0 \\ 0 & * & * & * \\ 0 & 0 & * & * \end{bmatrix}.$$

where the asterisk denotes elements yet to be determined, and proceed to construct a P and Q, if necessary, so that the elements e_{22} and g_{22} of PE and GQ, respectively, are both nonzero. (Note that if this is not possible, the factorization is complete.) Now compute the remaining elements in the second column of PE and the second row of GQ, and continue.

c. Compute A^+.

3.4 Let $A = [B,R]$ be any m by n upper trapezoidal matrix with $n \geq m + 2$ and let z_1 be any nonnull vector in $N(A)$. Show that Gaussian elimination, together with a permutation matrix, Q, can be used to reduce the matrix

$$\begin{bmatrix} A \\ z_1^H \end{bmatrix}$$

to an upper trapezoidal matrix S, say. Prove now that if z_2 is any vector such that $Sz_2 = 0$, then $Q^H z_2 \in N(A)$ and $(z_1, Q^H z_2) = 0$.

3.5 Apply the procedure in Exercise 3.4 to construct an orthonormal basis for $N(A)$ in Exercise 3.3.

3.6 Show that the two forms for $[A_{k-1}, a_k]^+$ in Theorem 6 can be written in a single expression as

$$[A_{k-1}, a_k]^+ = \begin{bmatrix} A_{k-1} - A_{k-1}^+ a_k d_k^H \\ \\ d_k^H \end{bmatrix}$$

where

$$d_k^H = c_k^+ + (1 - c_k^+ c_k)(1+\gamma_k)^{-1} a_k^H A_{k-1}^{H+} A_{k-1}^+.$$

3.7 Let $[A_{k-1}, a_k]^+$ be partitioned as

$$[A_{k-1}, a_k]^+ = \begin{bmatrix} G_{k-1} \\ \\ d_k^H \end{bmatrix}$$

with d_k^H a row vector

a. Show that

$$A_{k-1}^+ = \begin{cases} G_{k-1}[I + (1 - d_k^H a_k)^{-1} a_k d_k^H], & \text{if } d_k^H a_k \neq 1, \\ \\ G_{k-1}(I - d_k d_k^+), & \text{if } d_k^H a_k = 1. \end{cases}$$

b. Construct A^+ if

$$A = \begin{bmatrix} 1 & 0 & -1 & 0 \\ 2 & 1 & 0 & 2 \\ 3 & 0 & 1 & 4 \end{bmatrix}.$$

Hint: See Exercise 3.3c.

3.8 For any product AB let $B_1 = A^+ AB$ and $A_1 = AB_1 B_1^+$. Then $(AB)^+ = (A_1 B_1)^+ = B_1^+ A_1^+$. Why does this expression reduce to Theorem 4 when AB is a full rank factorization?

*3.9 a. Use Exercise 1.3 to prove that if A is any normal matrix,

$$A^+ = \sum \frac{1}{\lambda_i} x x_i^H$$

where Σ^1 indicates that the sum is taken over indices i with eigenvalues $\lambda_i \neq 0$.

b.   Prove that if A is normal, $(A^n)^+ = (A^+)^n$ for all $n \geq 1$.

c.   If $\lambda_i = 2i-2$, $i = 1,2,3$, and

$$x_1 = \frac{1}{\sqrt{2}}\begin{bmatrix} 1 \\ 0 \\ -1 \end{bmatrix}, \quad x_2 = 1/3\begin{bmatrix} 2 \\ 1 \\ 2 \end{bmatrix}, \quad x_3 = \frac{1}{3\sqrt{2}}\begin{bmatrix} 1 \\ -4 \\ 1 \end{bmatrix},$$

construct the Moore-Penrose inverse of the matrix, A, for which $Ax_i = \lambda_i x_i$, $i = 1,2,3$.

3.2   Applications with Matrices of Special Structure

For many applications of mathematics it is required to solve systems of equations Ax = b in which A or b or both A and b have some special structure resulting from the physical considerations of the particular problem. In some cases this special structure is such that we can obtain information concerning the set of all solutions. For example, the explicit form for all solutions of the equations

$$x_i + x_{n+1} = b_i, \quad i = 1,\ldots,n,$$

given in Exercise 2.11, was obtained using the Moore-Penrose inverse of the matrix [I,u] from Exercise 2.10 where u is the n-tuple with each element equal to unity. In this section we introduce the concept of the Kronecker product of matrices which can be used to characterize all solutions of certain classes of problems that occur in the design of experiments and in linear programming.

DEFINITION 2:   For any m by n matrix, P, and s by t matrix, $Q = (q_{k\ell})$, the *Kronecker product* of P and Q is the ms by nt matrix, $P \times Q$, of the form

$$P \times Q = (q_{k\ell}P).$$

It should be noted in Definition 2 that if $P = (p_{ij})$ and $Q = (q_{k\ell})$, then $P \times Q$ is obtained by replacing each element $q_{k\ell}$ by the matrix $q_{k\ell}P$, whereas $Q \times P$ is obtained by replacing each element p_{ij} by the matrix $p_{ij}Q$. Consequently $P \times Q$ and $Q \times P$ differ only in the order in which rows and columns appear, and there exist permutation matrices R and S, say, such that $Q \times P = R[P \times Q]S$. (We remark also that some authors, for example, Thrall and Tornheim [13], define the Kronecker product of P and Q alternately as $P \times Q = (p_{ij}Q)$, that is, our $Q \times P$. In view of the discussion in Section 3.1 of the Moore-Penrose inverses of matrices A and $\tilde{A}$, where $\tilde{A}$ is obtained by permuting rows of A, columns of A or both, each of the following results obtained using the form for $P \times Q$ in Definition 2 has a corresponding dual if the alternate definition is employed.)

Example 3.4

If

$$P = \begin{bmatrix} 1 & 2 \\ 3 & 0 \end{bmatrix}, \ Q = \begin{bmatrix} 0 & 4 & -1 \\ 2 & i & 3 \end{bmatrix},$$

then

$$P \times Q = \begin{bmatrix} 0 & 0 & 4 & 8 & -1 & -2 \\ 0 & 0 & 12 & 0 & -3 & 0 \\ 2 & 4 & i & 2i & 3 & 6 \\ 6 & 0 & 3i & 0 & 9 & 0 \end{bmatrix}.$$

and

$$Q \times P = \begin{bmatrix} 0 & 4 & -1 & 0 & 8 & -2 \\ 2 & i & 3 & 4 & 2i & 6 \\ 0 & 12 & -3 & 0 & 0 & 0 \\ 6 & 3i & 9 & 0 & 0 & 0 \end{bmatrix}.$$

Given any Kronecker product $P \times Q$, it follows from Definition 2 that

(3.6) $[P \times Q]^H = (\bar{q}_{\ell k}P^H) = P^H \times Q^H.$

-41-

Also, for any matrices R and S = $(s_{k\ell})$ with the products PR and QS defined, the product $[P \times Q][R \times S]$ is defined, and we have by use of block multiplication that

$$[P \times Q][R \times S] = \begin{bmatrix} q_{11}P & \cdots & q_{1n}P \\ \cdot & & \cdot \\ \cdot & & \cdot \\ \cdot & & \cdot \\ q_{m1}P & \cdots & q_{mn}P \end{bmatrix} \begin{bmatrix} s_{11}R & \cdots & s_{1t}R \\ \cdot & & \cdot \\ \cdot & & \cdot \\ \cdot & & \cdot \\ s_{n1}R & \cdots & s_{nt}R \end{bmatrix}$$

$$= \begin{bmatrix} \sum_{j=1}^{n} q_{1j}s_{ji}PR & \cdots & \sum_{j=1}^{n} q_{1j}s_{jt}PR \\ \cdot & & \cdot \\ \cdot & & \cdot \\ \cdot & & \cdot \\ \sum_{j=1}^{n} q_{mj}s_{j1}PR & \cdots & \sum_{j=1}^{n} q_{mj}s_{jt}PR \end{bmatrix}.$$

Therefore,

(3.7) $[P \times Q][R \times S] = PR \times QS.$

The following lemma can be established simply by combining the relationships in (3.6) and (3.7) to show that the defining equations in (2.2) are satisfied.

LEMMA 7:   For any matrices P and Q, $[P \times Q]^{+} = P^{+} \times Q^{+}.$ ∎

Example 3.5

To construct A^{+} if

$$A = \begin{bmatrix} 2 & 0 & 1 & 4 & 0 & 2 \\ 2 & 3 & -1 & 4 & 6 & -2 \\ 6 & 0 & 3 & 8 & 0 & 4 \\ 6 & 9 & -3 & 8 & 12 & -4 \end{bmatrix},$$

observe that $A = P \times Q$, where

$$P = \begin{bmatrix} 2 & 0 & 1 \\ 2 & \cdot 3 & -1 \end{bmatrix}, \quad Q = \begin{bmatrix} 1 & 2 \\ 3 & 4 \end{bmatrix}.$$

Then we have

$$Q^+ = Q^{-1} = -1/2 \begin{bmatrix} 4 & -2 \\ -3 & 1 \end{bmatrix}, \quad P^+ = \frac{1}{61} \begin{bmatrix} 22 & 4 \\ -9 & 15 \\ 17 & -8 \end{bmatrix}$$

so that

$$A^+ = P^+ \, X \, Q^+ = -\frac{1}{122} \begin{bmatrix} 88 & 16 & -44 & -8 \\ -36 & 60 & 18 & -30 \\ 68 & -32 & -34 & 16 \\ -66 & -12 & 22 & 4 \\ 27 & -45 & -9 & 15 \\ 51 & 24 & 17 & -8 \end{bmatrix}.$$

Example 3.6

To construct A^+ if

$$A = \begin{bmatrix} 1 & 1 & 0 & 0 & 1 & 0 & 1 & 0 & 0 & 0 & 0 & 0 \\ 1 & 0 & 1 & 0 & 1 & 0 & 0 & 1 & 0 & 0 & 0 & 0 \\ 1 & 0 & 0 & 1 & 1 & 0 & 0 & 0 & 1 & 0 & 0 & 0 \\ 1 & 1 & 0 & 0 & 0 & 1 & 0 & 0 & 0 & 1 & 0 & 0 \\ 1 & 0 & 1 & 0 & 0 & 1 & 0 & 0 & 0 & 0 & 1 & 0 \\ 1 & 0 & 0 & 1 & 0 & 1 & 0 & 0 & 0 & 0 & 0 & 1 \end{bmatrix}$$

observe first that if u_i denotes the vector with all i elements each equal to unity, then A can be written in partitioned form as

$$(3.8) \qquad A = \begin{bmatrix} u_3 & I_3 & u_3 & 0 & I_3 & 0 \\ u_3 & I_3 & 0 & u_3 & 0 & I_3 \end{bmatrix}.$$

Whereupon, the first two columns of A in (3.8) can be written as the Kronecker product $[u_3, I_3] \, X \, u_2$. Next observe that permuting columns of A to form

$$\tilde{A} = \begin{bmatrix} u_3 & I_3 & u_3 & I_3 & 0 & 0 \\ u_3 & I_3 & 0 & 0 & u_3 & I_3 \end{bmatrix}$$

the last four columns of $\tilde{A}$ become $[u_3, I_3] \, X \, I_2$. Therefore

-43-

$\tilde{A}$ can be written as

$$\tilde{A} = [u_3, I_3] \, X \, [u_2, I_2],$$

and thus

$$(3.9) \qquad \tilde{A}^+ = [u_3, I_3]^+ \, X \, [u_2, I_2]^+.$$

Permuting rows of $[I,u]^+$ in Exercise 2.10 now gives

$$[u_i, I_i]^+ = \frac{1}{i+1} \begin{bmatrix} u_i^H \\ (i+1)I_i - u_i u_i^H \end{bmatrix}$$

so that (3.9) becomes

$$(3.10) \qquad \tilde{A} = \frac{1}{12} \begin{bmatrix} u_3^H \\ 4I_3 - u_3 u_3^H \end{bmatrix} X \begin{bmatrix} u_2^H \\ 3I_2 - u_2 u_2^H \end{bmatrix}.$$

Substituting numerical values into (3.10), a suitable permutation of rows of $\tilde{A}^+$ yields A^+.

Matrices, A, as in Example 3.6, with elements zero or one occur frequently in the statistical design of experiments, and the technique of introducing Kronecker products can often be used to construct A^+, and thus all solutions of systems of equations Ax = b by use of Exercise 2.21. The additional restrictions on Ax = b to obtain solutions with particular properties can then be formulated in terms of conditions on N(A) or, equivalently, $I - A^+A$. (See Exercise 3.11.)

That Kronecker products can be combined with forms for Moore-Penrose inverses of partitioned matrices to construct A^+ for other classes of structured matrices is shown by the representation in Theorem 8.

THEOREM 8: Let W be any m by n matrix, and for any positive integer p let $G_p = (pI_n + W^H W)^{-1}$. Then

$$(3.11) \qquad \begin{bmatrix} I_n \mathbin{\chi} u_p^{H} \\ W \mathbin{\chi} I_p \end{bmatrix}^+ = [G_p \mathbin{\chi} u_p, W^+ \mathbin{\chi} I_p - G_p W^+ \mathbin{\chi} u_p u_p^{H}].$$

Proof: Observe first that with

$$(pI_n + W^H W)(I_n - W^+ W) = p(I_n - W^+ W) = (I_n - W^+ W)(pI_n + W^H W)$$

then

$$(3.12) \qquad I_n - W^+ W = pG_p(I_n - W^+ W) = p(I_n - W^+ W)G_p.$$

Also, observe that with $(pI_n + W^H W)W^+ = pW^+ + W^H$, the relation

$$(3.13) \qquad W^+ = pG_p W^+ + G_p W^H,$$

together with the fact that G_p is Hermitian, implies

$$WG_p W^+ = \frac{1}{p}(WW^+ - WG_p W^H)$$

is Hermitian.

Let

$$A = \begin{bmatrix} I_n \mathbin{\chi} u_p^{H} \\ W \mathbin{\chi} I_p \end{bmatrix},$$

and let

$$X = [G_p \mathbin{\chi} u_p, W^+ \mathbin{\chi} I_p - G_p W^+ \mathbin{\chi} u_p u_p^{H}].$$

Then it follows from (3.12) and (3.7) that

$$XA = G_p \mathbin{\chi} u_p u_p^{H} + W^+ W \mathbin{\chi} I_p - G_p W^+ W \mathbin{\chi} u_p u_p^{H}$$

$$= G_p(I_n - W^+ W) \mathbin{\chi} u_p u_p^{H} + W^+ W \mathbin{\chi} I_p$$

$$= \frac{1}{p}(I_n - W^+ W) \mathbin{\chi} u_p u_p^{H} + W^+ W \mathbin{\chi} I_p$$

is Hermitian. Also, with $u_p^{H} u_p = p$,

$$AXA = \begin{bmatrix} \frac{1}{p}(I_n - W^+W)\, X\, pu_p^H + W^+W\, X\, u_p^H \\[2ex] \frac{1}{p}W(I_n - W^+W)\, X\, u_p u_p^H + WW^+W\, X\, I_p \end{bmatrix} = \begin{bmatrix} I_p\, X\, u_p^H \\[2ex] W\, X\, I_p \end{bmatrix} = A.$$

Continuing, we have

$$XA(G_p\, X\, u_p) = \frac{1}{p}(I_n - W^+W)G_p\, X\, pu_p + W^+WG_p\, X\, u_p = G_p\, X\, u_p,$$

$$XA(W^+\, X\, I_p) = \frac{1}{p}(I_n - W^+W)W^+\, X\, u_p u_p^H + W^+WW^+\, X\, I_p = W^+\, X\, I_p,$$

and

$$XA(G_pW^+\, X\, u_p u_p^H) = \frac{1}{p}(I_n - W^+W)G_pW^+\, X\, pu_p u_p^H + W^+WG_pW^+\, X\, u_p u_p^H$$

$$= GW^+\, X\, u_p u_p^H.$$

Hence, XAX = X. Finally, forming AX gives

$$AX = \begin{bmatrix} pG_p & W^+\, X\, u_p^H - G_pW^+\, X\, pu_p^H \\[2ex] WG_p\, X\, u_p & WW^+\, X\, I_p - WG_pW^+\, X\, u_p u_p^H \end{bmatrix}.$$

Now

$$W^+\, X\, u_p^H - G_pW^+\, X\, pu_p^H = G_pW^H\, X\, u_p^H,$$

by (3.13), which, with G_p and WG_pW^+ Hermitian, implies $(AX)^H = AX$.

Having shown that A and X satisfy the equations in (2.2), then $X = A^+$ which establishes (3.11). ∎

Example 3.7

If $p = 3$, then for any m by n matrix, W,

$$\begin{bmatrix} I_n\, X\, u_3^H \\[2ex] W\, X\, I_3 \end{bmatrix} = \begin{bmatrix} I_n & I_n & I_n \\ W & 0 & 0 \\ 0 & W & 0 \\ 0 & 0 & W \end{bmatrix}$$

and

$$
\begin{bmatrix} I_n \chi u_3^H \\ \\ W \chi I_3 \end{bmatrix}^+ = \begin{bmatrix} G_3 & W^+ - G_3 W^+ & -G_3 W^+ & -G_3 W^+ \\ G_3 & -G_3 W^+ & W^+ - G_3 W^+ & -G_3 W^+ \\ G_3 & -G_3 W^+ & -G_3 W^+ & W^+ - G_3 W^+ \end{bmatrix},
$$

where $G_3 = (3I + W^H W)^{-1}$.

Suppose now that we let $T = T(p,W)$ denote the matrix in Theorem 8 which is completely determined by p and the sub-matrix W, that is,

$$
T = T(p,W) = \begin{bmatrix} I_n \chi u_p^H \\ \\ W \chi I_p \end{bmatrix}.
$$

Now given a system of equations $Tx = b$ with W m by n of rank r, $0 < r \le n$, and p any positive integer it follows that if we partition x and b as

$$
x = \begin{bmatrix} x^{(1)} \\ \cdot \\ \cdot \\ \cdot \\ x^{(p)} \end{bmatrix}, \quad b = \begin{bmatrix} b^{(0)} \\ \cdot \\ \cdot \\ \cdot \\ b^{(p)} \end{bmatrix}
$$

with $x^{(1)}, \ldots, x^{(p)}$ and $b^{(0)}$ n-tuples and $b^{(1)}, \ldots, b^{(p)}$ m-tuples, then x is a solution if and only if

$$
(3.14) \qquad \sum_{j=1}^{P} x^{(j)} = b^{(0)}
$$

and

$$
(3.15) \qquad Wx^{(j)} = b^{(j)}, \quad j = 1, \ldots, p.
$$

In other words, each $x^{(j)}$ must be a solution of m equations in n unknowns, subject to the condition that the sum of the solutions is equal to $b^{(0)}$. These characterizations are further explored for the general case of an arbitrary matrix, W, in Exercises 3.16 and 3.17 and for an important special case in Exercises 3.18 and 3.19.

3.10 Complete the numerical construction of A^+ in Example 3.6 and verify that A and A^+ satisfy the defining equations in (2.2).

3.11 Matrices of the form $[u_p, I_p]$ and, more generally, Kronecker products such as A in Example 3.6 in which at least one of the matrices has this form occur frequently in statistical design of experiments [1][4]. For example, suppose it is required to examine the effect of p different fertilizers on soy bean yield. One approach to this problem is to divide a field into pq subsections (called plots), randomly assign each of the p type of fertilizers to q plots, and measure the yield from each. Neglecting other factors which may effect yield, a model for this experiment has the form

$$(3.16) \quad y_{ij} = m + t_i + e_{ij}$$

where y_{ij} is the yield of the jth plot to which fertilizer i has been applied, m is an estimate of an overall "main" effect, t_i is an estimate of the effect of the particular fertilizer treatment and e_{ij} is the experimental error associated with the particular plot. The question now is to determine m and $t_1, \ldots, t_p$ to minimize the sum of squares of experimental error, that is,

$$\sum_{i=1}^{p} \sum_{j=1}^{q} e_{ij}^2.$$

a. If y and e denote the vectors

$$y = (y_{11}, \ldots, y_{p1}, y_{12}, \ldots, y_{p2}, \ldots, y_{1q}, \ldots, y_{pq})^H$$

and

$$e = (e_{11}, \ldots, e_{p1}, e_{12}, \ldots, e_{p2}, \ldots, e_{1q}, \ldots, e_{pq})^H,$$

show that data for the model in (3.16) can be represented as

$$(3.17) \quad y = Ax + e,$$

where $x = (m, t_1, \ldots, t_p)^H$ and $A = [u_p, I_p] \times u_q$.

b. Show that $r(A) = p$, and construct the minimal norm solution $\hat{x} = A^{+}y$, to (3.17).

c. For statistical applications it is also assumed that

$$\sum_{i=1}^{p} t_i = 0.$$

Starting with $\hat{x}$ from 3.11b above, construct that solution $\overset{\approx}{x}$, say, for which this additional condition holds.

3.12 (Continuation): Given the experimental situation described in Exercise 3.11, it is sometimes assumed that there is another effect, called a block effect, present. In this case one of two models is assumed: First, if there is no interaction between the treatment and block effect, then

$$(3.18) \quad y_{ij} = m + t_i + b_j + e_{ij},$$

whereas if there is an assumed interaction between the treatment and block effect, then

$$(3.19) \quad y_{ij} = m + t_i + b_j + (tb)_{ij} + e_{ij},$$

where y_{ij}, m and t_i have the same meaning as in (3.16), $b_j, j=1,\ldots,q$, designate block effects, and $(tb)_{ij}, i=1,\ldots,p$ and $j=1,\ldots,q$, designate the effect of the interaction between treatment i and block j.

a. Using the notation for y and e from Exercise 3.11, show that the data for the model in (3.18) can be represented as

$$y = A_1 x + e,$$

where now $x = (m, t_1, \ldots, t_p, b_1, \ldots, b_q)^H$ and

$$A_1 = \left[[u_p \,\mathsf{X}\, I_p] \,\mathsf{X}\, u_q, u_p \,\mathsf{X}\, I_q \right].$$

b. Show that the data for the model in (3.19) can be represented as

$$y = A_2 x + b$$

where

$$x = (m, t_1, \ldots, t_p, b_1, \ldots, b_q, (tb)_{11}, \ldots, (tb)_{1q}, \ldots, (tb)_{p1}, \ldots, (tb)_{pq})^H$$

and $A_2 = \left[[u_p, I_p] \, X u_q , u_p \, X I_q , I_p \, X I_q \right]$.

c. Use the procedure of Example 3.6 to construct A_2^+ and thus the solution $\hat{x} = A_2^+ y$. (For statistical applications the model in (3.19) is not meaningful unless there is more than one observation for each pair of indices i and j, that is, a model of the form

$$y_{ijk} = m + t_i + b_j + (tb)_{ij} + e_{ijk}$$

where $k = 1,\ldots,r$. In this case the unique solution is obtained by assuming

$$\sum_{i=1}^{p} t_i = \sum_{j=1}^{q} b_j = 0 \quad \text{and also} \quad \sum_{i=1}^{p} (tb)_{ij} = \sum_{j=1}^{q} (tb)_{ij} = 0$$

for all i and j. Note, in addition, that the construction of A_1^+ in 3.12a above is somewhat more complicated, but can be formed using related techniques. The particular solution used for statistical applications in this case assumes that

$$\sum_{i=1}^{p} t_i = \sum_{j=1}^{q} b_j = 0.)$$

*3.13 Show that if P and Q are any square matrices with x an eigenvector of P corresponding to eigenvalue λ and y an eigenvector of Q corresponding to eigenvalue μ, then $x \, X \, y$ is an eigenvector of $P \, X \, Q$ corresponding to eigenvalue $\lambda\mu$.

3.14 a. Show that for any p and W, $I - T^+ T = (I_n - W^+ W) \, X \, (I_p - u_p u_p^+)$.

   *b. Construct a complete orthonormal set of eigenvectors for $I - T^+ T$.

3.15 Prove directly that for any m by n matrix W of rank r and any p, rank $(T) = n + r(p-1)$.

3.16 For any system of equations $Tx = b$ with x and b partitioned to give (3.14) and (3.15), let X and B denote the matrices
$$X = [x^{(1)}, \ldots, x^{(p)}], \quad B = [b^{(1)}, \ldots, b^{(p)}].$$

   a. Prove that $Tx = b$ if and only if there exists a matrix X such that

(3.20) $Xu_p = b^{(o)}$

and

(3.21) $WX = B.$

b. Prove that a necessary condition for a solution, X, to (3.20)
and (3.21) to exist is $WW^+B = B$ and $Wb^{(o)} = Bu_p$.

*3.17 (Continuation): For any eigenvector $z_i \times y_j$ of $I_{np} - T^+T$, where
y_j has components $y_{j1}, \ldots, y_{jp}$, let Z_{ij} denote the matrix

$$Z_{ij} = [y_{j1}z_i, \ldots, y_{jp}z_i].$$

Show that $z_i \times y_j$ corresponds to eigenvalue $\lambda_i \gamma_j = 1$ if and only if

(3.22) $Z_{ij}u_p = 0$

and

(3.23) $WZ_{ij} = 0.$

3.18 The transportation problem in linear programming is an example of
a problem in which it is required to solve a system of equations
$Tx = b$. This famous problem can be stated as follows: Consider
a company with n plants which produce $a_1, \ldots, a_n$ units, respectively,
of a given product in some time period. This company has p
distributors which require $b_1, \ldots, b_p$ units, respectively, of the
product in the same time period, where

$$\sum_{i=1}^{n} a_i = \sum_{j=1}^{p} b_j.$$

If there is a unit cost c_{ij} for shipping from plant i to distributor
j, $i=1, \ldots, n$ and $j=1, \ldots, p$, then how should the shipments be allo-
cated in order to minimize total transportation cost? This problem
can be illustrated in a schematic form (called a tableau) as shown
in Figure 6 where $0_1, \ldots, 0_n$ designate origins of shipment (plants),
$D_1, \ldots, D_p$ designate destinations (distributors) and for each i and
j, x_{ij} denotes the number of units to be shipped from 0_i to D_j.

The problem now is to determine the x_{ij}, $i=1, \ldots, n$ and $j=1, \ldots, p$
to minimize the total shipping cost.

-51-

	D_1	. . .	D_j	. . .	D_p	
O_1	c_{11} x_{11}	. . .	c_{1j} x_{1j}	. . .	c_{1p} x_{1p}	a_1
⋮	⋮	. . .	⋮	. . .	⋮	⋮
O_i	c_{i1} x_{i1}	. . .	c_{ij} x_{ij}	. . .	c_{ip} x_{ip}	a_i
⋮	⋮	. . .	⋮	. . .	⋮	⋮
O_n	c_{n1} x_{n1}	. . .	c_{nj} x_{nj}	. . .	c_{np} x_{np}	a_n
	b_1	. . .	b_j	. . .	b_p	$\sum_i a_i = \sum_j b_j$

Figure 6. The transportation problem tableau.

$$\sum_{i=1}^{n} \sum_{j=1}^{p} c_{ij}x_{ij}, \tag{3.24}$$

subject to the conditions that

$$\sum_{j=1}^{p} x_{ij} = a_i, \quad i = 1,\ldots,n, \tag{3.25}$$

and

$$\sum_{i=1}^{n} x_{ij} = b_j, \quad j = 1,\ldots,p. \tag{3.26}$$

Also, we must have $x_{ij} \geq 0$, for all i and j, and, assuming fractional units cannot be manufactured or shipped, all a_i, b_j and x_{ij}

integers. (This last requirement that the x_{ij} are integers follows automatically when the a_i and b_j are [6].)

a. Show that if the x_{ij} in Figure 6 are elements of an n by p matrix X, and if $b^{(o)} = [a_1, \ldots, a_n]^H$, then the conditions in (3.25) can be written as $Xu_p = b^{(o)}$ and the conditions in (3.26) become

$$u_n^H X = [b_1, \ldots, b_p].$$

Therefore, (3.25) and (3.26) together imply that any set of numbers x_{ij} which satisfy the row and column requirements of the tableau is a solution of $Tx = b$ where $T = T(p, u_n^H)$ and $b = [a_1, \ldots, a_n, b_1, \ldots, b_p]^H$.

b. Prove that if $T = T(p, u_n^H)$, then

$$T^+ = \left[\left(I_n - \frac{1}{n+p} u_n u_n^H \right) X u_p^+, u_n^+ X \left(I_p - \frac{1}{n+p} u_p u_p^H \right) \right].$$

Moreover, show that if $\hat{x}_{ij}$ is the element in row i and column j of the tableau form of $\hat{x} = T^+ b$, then

$$\hat{x}_{ij} = \frac{1}{p} a_i + \frac{1}{n} b_j - \frac{1}{np} \sum_{i=1}^{n} a_i$$

for $i = 1, \ldots, n$ and $j = 1, \ldots, p$.

*c. Show that rank $(T) = n + p - 1$ when $W = u_n^H$, and thus rank $(I_{np} - T^+ T) = (n-1)(p-1)$. Also, construct a complete othonormal set of eigenvectors of $I_{np} - T^+ T$, and show that $z_i X y_j$ is an eigenvector corresponding to eigenvalue $\lambda_i \gamma_j = 1$ if and only if all row sums and column sums in the tableau form are zero.

d. The vector $g = (I_{np} - T^+ T)c$ is called the gradient of the inner product

$$(c, x) = \sum_{i=1}^{n} \sum_{j=1}^{p} c_{ij} x_{ij}$$

in (3.24). Show that the elements, g_{ij} in the tableau form for g can be written as

$$g_{ij} = c_{ij} - \frac{1}{n}\sum_{i=1}^{n}c_{ij} - \frac{1}{p}\sum_{j=1}^{p}c_{ij} + \frac{1}{np}\sum_{i=1}^{n}\sum_{j=1}^{p}c_{ij}$$

for $i = 1,\ldots,n$ and $j=1,\ldots,p$.

3.19 (Continuation): The transportation problem has been generalized in a number of different ways, and one of these extensions follows directly using matrices of the form $T = T(p,W)$. Suppose that we are given q transportation problems, each with n origins and p destinations, and let a_{ik}, b_{jk}, c_{ijk} and x_{ijk} be the row sums, column sums, costs and variables, respectively, associated with the kth tableau, $k=1,\ldots,q$. A "three-dimensional" transportation problem is now obtained by adding the conditions that

(3.27) $$\sum_{k=1}^{q} x_{ijk} = d_{ij}$$

for $i=1,\ldots,n$ and $j=1,\ldots,p$, where $d_{11},\ldots,d_{np}$ are given positive integers. (The choice of nomenclature "three-dimensional is apparent by noting that if the tableaus are stacked to form a parallelopiped with q layers each with np cells, then (3.27) simply implies np conditions that must be satisfied when the x_{ijk} are summed in the vertical direction as shown in Figure 7, where only the row, column and vertical sum requirements are indicated.)

a. Show that the conditions

$$\sum_{i=1}^{n} a_{ik} = \sum_{j=1}^{p} b_{jk}, \quad k=1,\ldots,q,$$

$$\sum_{k=1}^{q} a_{ik} = \sum_{j=1}^{p} d_{ij}, \quad i=1,\ldots,n,$$

$$\sum_{k=1}^{q} b_{jk} = \sum_{i=1}^{n} d_{ij}, \quad j=1,\ldots,p,$$

are necessary in order for a three-dimensional transportation problems to have a solution.

b. Show that the conditions which the x_{ijk} must satisfy if there is a solution can be written as $Tx = b$ where $T = T(q,W)$ with

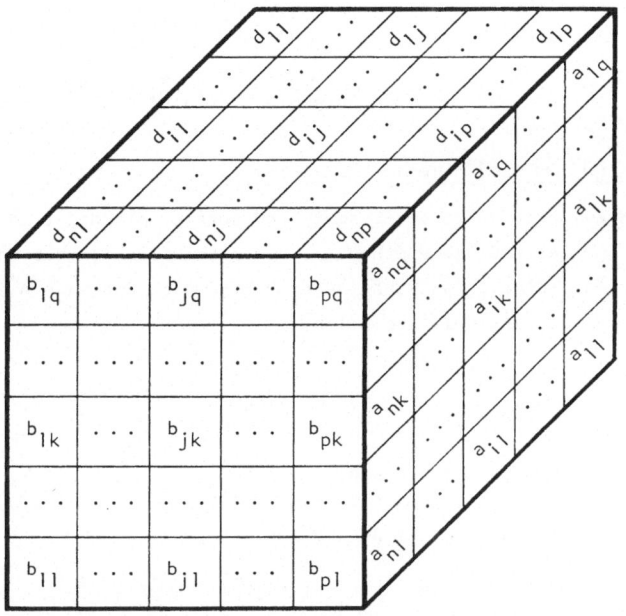

Figure 7. The parallelopiped requirements for the
tableau of a three-dimensional transportation problem.

$W = T_{np} = T(p, u_n^H)$ the matrix for the "two-dimensional transportation problem in Exercise 3.18 and a suitable vector b.

c. Show that

$$G_q = \left[qI_{np} + T_{np}{}^H T_{np} \right]^{-1}$$

$$= \left[\frac{1}{q}\left(I_p - \frac{1}{p+q} u_p u_p{}^H \right) \chi I_n \right]$$

$$- \left[\frac{1}{q(n+q)}\left(I_p - \frac{n+p+2q}{(p+q)(n+p+q)} u_p u_p{}^H \right) \chi u_n u_n{}^H \right],$$

and that $G_q T_{np}{}^+ = [U, V]$ where

$$U = \frac{1}{n(n+q)}\left(I_p - \frac{2n+p+q}{(p+q)(n+p+q)} u_p u_p{}^H \right) \chi u_n$$

-55-

and

$$V = u_p X \frac{1}{n(p+q)} \left(I_n - \frac{n + 2p + q}{(n+p)(n+p+q)} u_n u_n^H \right).$$

3.3 Miscellaneous Exercises

3.20 Prove that a necessary and sufficient condition that the equations $AX = C$, $XB = D$ have a common solution is that each equation has a solution and that $AD = CB$, in which case $X = A^+C + DB^+ - A^+ADB^+$ is a particular solution.

3.21 Prove Lemma 7.

3.22 Prove that

$$\left(I + B^HB - B^+B \right) \left(I + B^+B^{H+} \right) = I + B^HB$$

for any matrix B, and that

$$\left(I + B^HB \right)^{-1} + \left(I + B^+B^{H+} \right)^{-1} = 2I - B^+B.$$

4

Drazin Inverses

4.1 The Drazin Inverse of a Square Matrix

In this section we consider another type of generalized inverse for square complex matrices. The inverse in Theorem 9, due to Drazin [3], has a variety of applications.

THEOREM 9: For any square matrix, A, there is a unique matrix X such that

(4.1) $A^k = A^{k+1}X$, for some positive integer k,

(4.2) $X^2A = X$,

(4.3) $AX = XA$.

Proof: Observe first that if A = 0 is the null matrix, then A and X = 0 satisfy (4.1), (4.2) and (4.3).

Suppose $A \neq 0$ is any n by n matrix. Then there exist scalars $d_1,\ldots,d_t$, not all zero, such that

$$\sum_{i=1}^{t} d_i A^i = 0,$$

where $t \leq n^2+1$ since the A^i can be viewed as vectors with n^2 elements. Let d_k be the first nonzero coefficient. Then we can write

(4.4) $A^k = A^{k+1}U,$

where

$$U = -\frac{1}{d_k}\left(\sum_{i=k+1}^{t} d_i A^{i-k-1}\right).$$

Since U is a polynomial in A, U and A commute. Also, multiplying both sides of (4.4) by AU gives

$$A^k = A^{k+2}U^2 = A^{k+3}U^3 = \ldots,$$

and thus

(4.5) $A^k = A^{k+m}U^m$

for all $m \geq 1$.

Let $X = A^k U^{k+1}$. Then for this choice of X,

$$A^{k+1}X = A^{2k+1}U^{k+1} = A^k$$

and

$$X^2 A = A^k U^{k+1} A^k U^{k+1} A = (A^{2k+1}U^{k+1})U^{k+1} = A^k U^{k+1} = X,$$

by use of (4.5). Also, X and A commute since U and A commute. Thus the conditions (4.1), (4.2) and (4.3) hold for this X.

To show that X is unique, suppose that Y is also a solution to (4.1), (4.2) and (4.3), where X corresponds to an exponent k_1 and Y corresponds to an exponent k_2 in (4.1). Let $\hat{k} = $ maximum (k_1, k_2). Then it follows using (4.1), (4.2), (4.3) and (4.5) that

$$X = X^2 A = X^3 A^2 = \ldots = X^{\hat{k}+1}A^{\hat{k}} = X^{\hat{k}+1}A^{\hat{k}+1}Y$$
$$= XAY = \ldots = XA^{\hat{k}+1}Y^{\hat{k}+1} = A^{\hat{k}}Y^{\hat{k}+1}$$
$$= \ldots = AY^2 = Y^2 A = Y$$

to establish uniqueness. ∎

We will call the unique matrix X in Theorem 9 the *Drazin inverse* of A and write X alternately as $X = A_d$. Also, we will call the smallest k such that (4.1) holds the *index* of A.

That A_d is a generalized inverse of A is apparent by noting that (4.1) holds with k = 1 when $X = A^{-1}$ exists and also (4.2) and (4.3) hold. Observe, moreover, that in general (4.1) can be rewritten as

$$(4.6) \qquad A^k XA = A^k$$

and (4.2) becomes XAX = X, by use of (4.3), so that the defining equations in Theorem 9 can be viewed as an alternative to those used for A^+ in which AXA = A is replaced by (4.6), (1.2) remains unchanged, and (1.3) and (1.4) are replaced by the condition in (4.3) that A and X commute. (Various relationships between A_d and A^+ will be explored in the exercises at the end of this section and in Section 4.3.)

As will be discussed following the proof of Lemma 10, full rank factorizations of A can be used effectively in the construction of A_d.

LEMMA 10: For any factorization A = BC, $A_d = B(CB)_d^2 C$.

Proof: Observe first that for any square matrix A and positive integers k, m and n, we have $A_d^m A^n = A_d^{m-n}$ if m > n and $A^{m+n} A_d^n = A^m$ if $m \geq k$ and A has index k.

Let k denote the larger of the index of BC and the index of CB. Then

$$A_d = (BC)_d = (BC)^{k+1}(BC)_d^{k+2} = B(CB)^k C(BC)_d^{k+2}$$

$$= B(CB)_d^{k+2}(CB)^{2k+2} C(BC)_d^{k+2}$$

$$= B(CB)_d^{k+2} C(BC)^{2k+2}(BC)_d^{k+2} = B(CB)_d^{k+2} C(BC)^k$$

$$= B(CB)_d^{k+2}(CB)^k C = B(CB)_d^2 C. \quad \blacksquare$$

Suppose now that $A = B_1 C_1$ is a full rank factorization where rank $(A) = r_1$. Forming the r_1 by r_1 matrix $C_1 B_1$, then either $C_1 B_1$ is nonsingular, or $C_1 B_1 = 0$, or rank $(C_1 B_1) = r_2$ where $0 < r_2 < r_1$. In the first case, with $C_1 B_1$ nonsingular, $(C_1 B_1)_d = (C_1 B_1)^{-1}$ so that $A_d = B_1 (C_1 B_1)^{-2} C_1$, by Lemma 10, where

$$(C_1 B_1)^{-2} = \left[(C_1 B_1)^{-1} \right]^2.$$

On the other hand, if $C_1 B_1 = 0$ then $(C_1 B_1)_d = 0$ and thus $A_d = 0$ by again using Lemma 10. Finally, if rank $(C_1 B_1) = r_2, 0 < r_2 < r_1$, then for any full rank factorization $C_1 B_1 = B_2 C_2$, we have

$$(C_1 B_1)_d = B_2 (C_2 B_2)_d^2 C_2$$

so that A_d in Lemma 10 becomes $A_d = B_1 B_2 (C_2 B_2)_d^3 C_2 C_1$. The same argument now applies to $C_2 B_2$, that is, either $C_2 B_2$ is nonsingular and

$$(C_2 B_2)_d^3 = (C_2 B_2)^{-3},$$

or $C_2 B_2 = 0$ and thus $A_d = 0$, or rank $(C_1 B_1) = r_3$ where $0 < r_3 < r_2$, and $C_2 B_2 = B_3 C_3$ is a full rank factorization to which Lemma 10 can be applied. Continuing in this manner with

$$\text{rank } (B_i C_i) \geq \text{rank } (C_i B_i) = \text{rank } (B_{i+1} C_{i+1}), \quad i = 1, 2, \ldots,$$

then either $B_m C_m = 0$ for some index m, and so $A_d = 0$, or rank $(B_m C_m) = \text{rank } (C_m B_m) > 0$ for some index m, in which case

$$(B_m C_m)_d = B_m (C_m B_m)^{-2} C_m$$

and thus

(4.7) $\qquad A_d = B_1 B_2 \cdots B_m (C_m B_m)^{-m-1} C_m C_{m-1} \cdots C_1$

in Lemma 10. Observe, moreover, that with $A = B_1 C_1$, $A^2 = B_1 C_1 B_1 C_1 = B_1 B_2 C_2 C_1, \ldots, A^m = B_1 B_2 \cdots B_m C_m C_{m-1} \cdots C_1$ and

(4.8) $\qquad A^{m+1} = B_1 B_2 \cdots B_m (C_m B_m) C_m C_{m-1} \cdots C_1$

we have either $A^m = A^{m+1} = 0$, and $A_d = 0$, or that A_d has the form in (4.7) where, since each B_i has full column rank and each C_i has full row rank,

$$B_{m-1}^+ \cdots B_1^+ A^m C_1^+ \cdots C_{m-1}^+ = B_m C_m$$

and

$$B_m^+ \cdots B_1^+ A^{m+1} C_1^+ \cdots C_m^+ = C_m B_m .$$

Therefore, in both cases we have rank (A^m) = rank (A^{m+1}). Furthermore, it follows in both cases that (4.1) holds for $k = m$ and does not hold for any $k < m$. That is to say, k in (4.1) is the smallest positive integer such that A^k and A^{k+1} have the same rank.

Example 4.1

If A is the singular matrix

$$A = \begin{bmatrix} 6 & 4 & 0 \\ 3 & 5 & -3 \\ 3 & 3 & -1 \end{bmatrix}$$

written as the full rank factorization

$$A = B_1 C_1 = \begin{bmatrix} 1 & 2 \\ 2 & 1 \\ 1 & 1 \end{bmatrix} \begin{bmatrix} 0 & 2 & -2 \\ 3 & 1 & 1 \end{bmatrix} ,$$

then

$$C_1 B_1 = \begin{bmatrix} 2 & 0 \\ 6 & 8 \end{bmatrix}$$

is nonsingular, so that A has index one, and

$$A_d = B_1(C_1B_1)^{-2}C_1 = \frac{1}{64}\begin{bmatrix} 6 & -26 & 30 \\ 3 & 35 & -33 \\ 3 & 3 & -1 \end{bmatrix}$$

Example 4.2

If A is the matrix

$$A = \begin{bmatrix} 7 & 0 & 0 & 0 \\ 0 & 0 & 1 & 2 \\ 0 & 0 & 0 & 3 \\ 0 & 0 & 0 & 0 \end{bmatrix}$$

with $A = B_1C_1$ the full rank factorization,

$$A = B_1C_1 = \begin{bmatrix} 7 & 0 & 0 \\ 0 & 1 & 0 \\ 0 & 1 & 1 \\ 0 & 0 & 0 \end{bmatrix}\begin{bmatrix} 1 & 0 & 0 & 0 \\ 0 & 0 & 1 & 2 \\ 0 & 0 & -1 & 1 \end{bmatrix}$$

where

$$C_1B_1 = \begin{bmatrix} 7 & 0 & 0 \\ 0 & 1 & 1 \\ 0 & -1 & -1 \end{bmatrix},$$

then rank $(C_1B_1) = 2$ and

$$C_1B_1 = B_2C_2 = \begin{bmatrix} 7 & 0 \\ 0 & 1 \\ 0 & -1 \end{bmatrix}\begin{bmatrix} 1 & 0 & 0 \\ 0 & 1 & 1 \end{bmatrix}$$

is a full rank factorization. Continuing,

$$C_2B_2 = \begin{bmatrix} 7 & 0 \\ 0 & 0 \end{bmatrix}$$

so that

$$C_2B_2 = B_3C_3 = \begin{bmatrix} 7 \\ 0 \end{bmatrix}\begin{bmatrix} 1 & 0 \end{bmatrix}$$

is a full rank factorization with $C_3B_3 = 7$. Hence A has index three and A_d becomes

$$A_d = B_1 B_2 B_3 (C_3 B_3)^{-4} C_3 C_2 C_1$$

$$= \frac{1}{2401} \begin{bmatrix} 343 \\ 0 \\ 0 \\ 0 \end{bmatrix} \begin{bmatrix} 1 & 0 & 0 & 0 \end{bmatrix} = \begin{bmatrix} \frac{1}{7} & 0 & 0 & 0 \\ 0 & 0 & 0 & 0 \\ 0 & 0 & 0 & 0 \\ 0 & 0 & 0 & 0 \end{bmatrix}.$$

For the special case of matrices with index one we have

(4.9) $AA_d A = A, \quad A_d AA_d = A_d, \quad AA_d = A_d A,$

so that

(4.10) $(A_d)_d = A$

by the duality in the roles of A and A_d. Conversely, if (4.10) holds, then the first and last relations in (4.9) follow from the defining relations in (4.2) and (4.3) applied to $(A_d)_d$ and A_d, and the second relation in (4.9) is simply (4.2) for A_d and A. Consequently, (4.10) holds if and only if A has index one. In this special case the Drazin inverse of A is frequently called the *group inverse* of A, and is designated alternately as $A^\#$. Thus $X = A^\#$, when it exists, is the unique solution of $AXA = A$, $XAX = A$ and $AX = XA$, and it follows from Lemma 10 that for any full rank factorization $A = BC$, $A^\# = B(CB)^{-2}C$.

Exercises

4.1   Compute A_d for the matrices

$$A_1 = \begin{bmatrix} 7 & 8 & 5 \\ 4 & 5 & 3 \\ 5 & 7 & 4 \end{bmatrix}, \quad A_2 = \begin{bmatrix} 0 & 2 & 3 \\ 0 & 0 & -1 \\ 0 & 0 & 0 \end{bmatrix}, \quad A_3 = \begin{bmatrix} 1 & 1 & 2 & -1 \\ 1 & 0 & 1 & 0 \\ 0 & 1 & 0 & 1 \\ -2 & -2 & -3 & 0 \end{bmatrix}.$$

4.2   Given any matrices B and C of the same size where B has full column rank, we will say that C is *alias* to B if $B^+ = (C^H B)^+ C^H$.

   a.   Prove that if C is alias to B, then $C^H B$ is nonsingular.

   b.   Show that the set of all matrices alias to B form an equivalence class.

c. Prove that $A_d = A^+$ if and only if C is alias to B for any full rank factorization $A = BC^H$.

d. Note, in particular, that $A_d = A^+$ when A is Hermitian. Prove this fact directly and also by using the result in 4.2c above.

4.3 Prove that $(A^H)_d = A_d^{\ H}$ and that $\left[(A_d)_d\right]_d = A_d$ for any matrix A.

4.4 Prove that $A_d = 0$ for any nilpotent matrix A.

4.5 Prove that $[P \times Q]_d = P_d \times Q_d$ for any square matrices P and Q. What is the index of $P \times Q$?

4.2 An Extension to Rectangular Matrices

The Drazin inverse of a matrix, A, as defined in Theorem 9, exists only if A is square, and an obvious question is how this definition can be extended to rectangular matrices. One approach to this problem is to observe that if B is a m by n with m > n, say, then B can be augmented by m-n columns of zeroes to form a square matrix A. Now forming A_d, we might then take those columns of A_d which correspond to the locations of columns of B in A as a definition of the "Drazin inverse" of B. As shown in the following example, however, the difficulty in this approach is that there are $\binom{m}{m-n}$ such matrices A, obtained by considering all possible arrangements of the n columns of B (taken without any permutations) and the m-n columns of zeroes, and that A_d can be different in each case.

Example 4.3

If

$$B = \begin{bmatrix} 1 & 2 \\ 0 & 1 \\ 3 & -1 \end{bmatrix} \quad \text{and}$$

$$A_1 = \begin{bmatrix} 0 & 1 & 2 \\ 0 & 0 & 1 \\ 0 & 3 & -1 \end{bmatrix}, \quad A_2 = \begin{bmatrix} 1 & 0 & 2 \\ 0 & 0 & 1 \\ 3 & 0 & -1 \end{bmatrix}, \quad A_3 = \begin{bmatrix} 1 & 2 & 0 \\ 0 & 1 & 0 \\ 3 & -1 & 0 \end{bmatrix},$$

then

$$(A_1)_d = \frac{1}{9}\begin{bmatrix} 0 & 10 & 7 \\ 0 & 3 & 3 \\ 0 & 9 & 0 \end{bmatrix}, \quad (A_2)_d = \frac{1}{7}\begin{bmatrix} 1 & 0 & 2 \\ 0 & 0 & 1 \\ 3 & 0 & -1 \end{bmatrix},$$

$$(A_3)_d = \begin{bmatrix} 1 & -2 & 0 \\ 0 & 1 & 0 \\ 3 & -13 & 0 \end{bmatrix}$$

are obtained by applying Lemma 10 to the matrices $A_i = BC_i$ where

$$C_1 = \begin{bmatrix} 0 & 1 & 0 \\ 0 & 0 & 1 \end{bmatrix}, \quad C_2 = \begin{bmatrix} 1 & 0 & 0 \\ 0 & 0 & 1 \end{bmatrix}, \quad C_3 = \begin{bmatrix} 1 & 0 & 0 \\ 0 & 1 & 0 \end{bmatrix}.$$

Observe in Example 4.3 that the nonzero columns of each matrix $(A_i)_d$ correspond to the product $B(C_iB)_d^{\ 2}$. Consequently, using the nonzero columns of $(A_i)_d$ to define the "Drazin inverse" of B implies that the resulting matrix is a function of C_i. That such matrices are uniquely determined by a set of defining equations and are special cases of a class of generalized inverses that can be constructed for any matrix B will be apparent from Theorem 11.

THEOREM 11: For any m by n matrix B and any n by m matrix W, there is a unique matrix X such that

(4.11) $(BW)^k = (BW)^{k+1}XW$, for some positive integer k,

(4.12) $XWBWX = X$,

(4.13) $BWX = XWB$.

Proof: Let $X = B(WB)_d^{\ 2}$. Then with $XW = B(WB)_d^{\ 2}W = (BW)_d$, by Lemma 10, (4.11) holds with k the index of BW. Also,

$$XWBWX = B(WB)_d^{\ 2}WBWB(WB)_d^{\ 2} = B(WB)_d^{\ 2} = X$$

and

$$BWX = BWB(WB)_d^{\ 2} = B(WB)_d^{\ 2}WB = XWB,$$

so that (4.12) and (4.13) hold.

To show that X is unique we can proceed as in the proof of Theorem 9. Thus, suppose X_1 and X_2 are solutions of (4.11), (4.12) and (4.13) corresponding to positive integers k_1 and k_2, respectively, in (4.11). Then with $\hat{k}$ = maximum (k_1, k_2), it follows that

$$X_1 = X_1 WBWX_1 = BWX_1 WX_1 = (BW)^2 (X_1 W)^2 X_1$$

$$= \ldots = (BW)^{\hat{k}} (X_1 W)^{\hat{k}} X_1 = (BW)^{\hat{k}+1} X_2 W (X_1 W)^{\hat{k}} X_1$$

$$= X_2 (WB)^{\hat{k}+1} W (X_1 W)^{\hat{k}} X_1 = X_2 WBW (BW)^{\hat{k}} (X_1 W)^{\hat{k}} X_1$$

$$= X_2 WBWX_1 .$$

Continuing in a similar manner with

$$X_2 = X_2 WBWX_2 = X_2 WX_2 WB = X_2 (WX_2)^2 (WB)^2$$

$$= \ldots = X_2 (WX_2)^{\hat{k}+1} (WB)^{\hat{k}+1} ,$$

then

$$X_2 WBWX_1 = X_2 (WX_2)^{\hat{k}+1} (WB)^{\hat{k}+1} WBWX_1$$

$$= X_2 (WX_2)^{\hat{k}+1} W (BW)^{\hat{k}+1} X_1 WB$$

$$= X_2 (WX_2)^{\hat{k}+1} W (BW)^{\hat{k}} B$$

$$= X_2 (WX_2)^{\hat{k}+1} (WB)^{\hat{k}+1} = X_2 .$$

Therefore, with $X_1 = X_2$, the solution to (4.11), (4.12) and (4.13) is unique. ∎

The unique matrix, X, in Theorem 11 will be called the *W-weighted Drazin inverse* of B and will be written alternately as $X = (B_W)_d$.

The choice of nomenclature W-weighted Drazin inverse of B is easily seen by noting that with $(B_W)_d = B(WB)_d^2$, then

$(B_W)_d = B_d$ when B is square and W is the identity matrix. Also, observe more generally that with B and $(B_W)_d$ of the same size and with W and WBW the size of B^H, the relation $BW(B_W)_d = (B_W)_d WB$ in (4.13) can be viewed as a generalized commutativity condition, and $(B_W)_d WBW(B_W)_d = (B_W)_d$ in (4.12) is analogous to (4.2) when written in the form XAX = X.

Example 4.4

If

$$B = \begin{bmatrix} 1 & 2 \\ 0 & 1 \\ 3 & -1 \end{bmatrix}$$

is the matrix in Example 4.3, and

$$W_1 = \begin{bmatrix} 1 & 2 & 4 \\ 3 & -1 & -2 \end{bmatrix}, \quad W_2 = \begin{bmatrix} 1 & 3 & 1 \\ 0 & 0 & 0 \end{bmatrix},$$

then

$$(B_{W_1})_d = \frac{1}{(91)^2} \begin{bmatrix} 169 & 338 \\ 60 & 169 \\ 87 & -169 \end{bmatrix}, \quad (B_{W_2})_d = \frac{1}{16} \begin{bmatrix} 1 & 1 \\ 0 & 0 \\ 3 & 3 \end{bmatrix}.$$

Exercises

4.6 Verify that B and $(B_W)_d$ satisfy the defining equations in Theorem 11 for $W = C_1$, C_2, C_3 in Example 4.3 and for $W = W_1$, W_2 in Example 4.4.

4.7 Prove that $E = E_d$ for any idempotent matrix E, and thus that $(B_W)_d = B^2 B_d$ when B is square and $W = B_d$. (Consequently, $(B_W)_d = B$ when $W = B_d$ and B has index one.)

4.8 Show that if W^H is any matrix alias to B, then $(B_W)_d = W^+(WB)^{-1}$.

4.9 Prove that $(B_W)_d = B^{H+}B^+B^{H+}$ for any matrix B when $W = B^H$. (Note that this result follows at once from Lemma 5(f) and Exercise 4.8 if B has full column rank, whereas Lemma 5(f) and Exercise 4.2d can be used for the general case.)

4.3 Expressions Relating A_d and A^+

It is an immediate consequence of Exercise 4.9 that if $W = B^H$ and so

$$(B_W)_d = B^{H+}B^+B^{H+},$$

then

$$B^+ = W(B_W)_d W.$$

Thus, using W-weighted Drazin inverses with $W = B^H$, $(B_W)_d$ and B^+ are related directly in terms of products of matrices which implies that $(B_W)_d$ and B^+ have the same rank. In contrast, for any square matrix, A, we have

$$\text{rank } (A_d) = \text{rank } (A^k) = \text{rank } (A^{k+1}),$$

with k the index of A, whereas rank $(A^+) = \text{rank } (A)$. Therefore, rank $(A_d) \leq \text{rank } (A^+)$ with equality holding if and only if A has a group inverse. The following result can be used to give a general expression for the Drazin inverse of a matrix, A, in terms of powers of A and a Moore-Penrose inverse.

THEOREM 12: For any square matrix A with index k,

$$(4.14) \qquad A_d = A^k Y A^k$$

for any matrix Y such that

$$(4.15) \qquad A^{2k+1} Y A^{2k+1} = A^{2k+1}.$$

Proof: Starting with the right-hand side of (4.14) we have

$$A^k Y A^k = A_d^{k+1} A^{2k+1} Y A^{2k+1} A_d^{k+1}$$

$$= A_d^{k+1} A^{2k+1} A_d^{k+1} = A_d^{2k+2} A^{2k+1} = A_d. \quad \blacksquare$$

Observe in (4.15) that one obvious choice of Y is $(A^{2k+1})^+$, and it then follows that A^ℓ, $(A^\ell)^+$ and A_d have the same rank for every positive integer $\ell \geq k$. In this case, various relationships among $A^\ell, (A^\ell)^+$ and A_d can be established. For example, it can be shown that for any $\ell \geq k$, there is a unique matrix X satisfying

(4.16) $A^\ell X A^\ell = A^\ell, \quad X A^\ell X = X$

and

(4.17) $(X A^\ell)^H = X A^\ell, \quad A^\ell X = A A_d.$

Dually, there is a unique matrix X satisfying (4.16) and

(4.18) $(A^\ell X)^H = A^\ell X, \quad X A^\ell = A A_d.$

The unique solutions of (4.16) and (4.17) and of (4.16) and (4.18) are called the *left* and *right power inverses* of A^ℓ, respectively, and are designated as $(A^\ell)_L$ and $(A^\ell)_R$. Moreover, it can be shown (Exercise 4.11) that

(4.19) $(A^\ell)_L = (A^\ell)^+ A A_d, \quad (A^\ell)_R = A_d A (A^\ell)^+,$

and (Exercise 4.14) that $(A^\ell)_L$ and $(A^\ell)_R$ can be computed using full rank factorizations.

Exercises

4.10 Show that if A and W satisfy $A^\ell W A^\ell = A^\ell$ and $(W A^\ell)^H = W A^\ell$ for any positive integer ℓ, then $W A^\ell = (A^\ell)^+ A^\ell$, and conversely. What is the dual form of this result for $A^\ell W$?

4.11 Prove that $(A^\ell)_L$ in (4.19) is the unique solution to (4.16) and (4.17).

4.12 Prove that for every $\ell \geq k$, $(A^\ell)^+ = (A^\ell)_L A^\ell (A^\ell)_R$ and $A_d = (A^\ell)_R A^{2\ell-1} (A^\ell)_L.$

4.13 Use a sequence of full rank factorizations

$$A = B_1 C_1, \quad A^2 = B_1 B_2 C_2 C_1, \ldots,$$

to show that $A^\ell (A^\ell)^+ = A^k (A^k)^+$ and $(A^\ell)^+ A^\ell = (A^k)^+ A^k$ for all $\ell \geq k$.

4.14 (Continuation): Show that

$$(A^{\ell})_L = \left(\prod_{i=1}^{k} C_{k+1-i}\right)^{+} (C_k B_k)^{-\ell} \left(\prod_{i=1}^{k} C_{k+1-i}\right),$$

and

$$(A^{\ell})_R = \left(\prod_{i=1}^{k} B_i\right) (C_k B_k)^{-\ell} \left(\prod_{i=1}^{k} B_i\right)^{+}.$$

4.15 Construct $(A^2)_L$ and $(A^2)_R$ for the matrix A in Example 4.1

4.16 Prove that if Ax = b is a consistent system of equations and if A has index one, then the general solution of $A^n x = b$, n = 1,2,..., can be written as $x = A_R^{\;n} b + (I-A_L A)y$ where y is arbitrary. (Note that this expression reduces to $x = A^{-n} b$ when A is nonsingular. The terminology "power inverse" of A was chosen since we use powers of A_R in a similar manner to obtain a particular solution of $A^n x = b$.)

4.4   Miscellaneous Exercises

4.17 Let B and W be any matrices, m by n and n by m, respectively, and let p be any positive integer.

   a.   Show that there is a unique matrix X such that

$$(BW)_d XW = (BW)_d^{\;p}, \quad BWX = XWB, \quad BW(BW)_d X = X,$$

   a unique matrix X such that

$$XW = BW(BW)_d^{\;p}, \quad WX = WB(WB)_d^{\;p}, \quad XW(BW)^{p-1} X = X,$$

   and that the unique X which satisfies both sets of equations is $X = B(WB)_d^{\;p}$.

   b.   Show that if $p \geq 1$, $q \geq -1$ and $r \geq 0$ are integers such that $q + 2r + 2 = p$, and if $(WB)^q = (WB)_d$ when q = -1, then

$$B(WB)_d^{\;p} = B(WB)^q [((WB)^r W)(B(WB)^q)]_d^{\;2}.$$

   (Consequently, the unique X in 4.17a is the $(WB)^r$ W-weighted Drazin inverse of $B(WB)^q$.)

4.18 Prove that if A and B are any matrices such that $A_d^{\;2} = B_d^{\;2}$, then $AA_d = BB_d$.

5

Other Generalized Inverses

5.1 Inverses That Are Not Unique

Given matrices A and X, subsets of the relations in
(1.1) to (1.5) other than those used to define A^+ and A_d
provide additional types of generalized inverses. Although
not unique, some of these generalized inverses exhibit the
essential properties of A^+ required in various applications.
For example, observe that only the condition $AXA = A$ was
needed to characterize consistent systems of equations
$Ax = b$ by the relation $AXb = b$ in (2.4). Moreover, if A
and X also satisfy $(XA)^H = XA$, then $XA = A^+A$, by Exercise
4.10, and with A^+b a particular solution of $Ax = b$, the
general solution in Exercise 2.21 can be written as

$$x = A^+b + (I-XA)y$$

with the orthogonal decomposition

$$||x||^2 = ||A^+b||^2 + ||(I-XA)y||^2.$$

(Note that this is an extension of the special case of matrices with full row rank used in the proof of Theorem 2.) In this section we consider relationships among certain of these generalized inverses in terms of full rank factorizations, and illustrate the construction of such inverses with numerical examples.

For any A and X such that AXA = A, rank (X) $\geq$ rank (A), whereas XAX = X implies rank (X) $\leq$ rank (A). The following lemma characterizes solutions of AXA = A and XAX = X in terms of group inverses.

LEMMA 13: For any full rank factorizations A = BC and X = YZ, AXA = A and XAX = X if and only if AX = $(BZ)^{\#}BZ$ and XA = $(YC)^{\#}YC$.

Proof: If A = BC and X = YZ are full rank factorizations where B is m by r, C is r by n, Y is n by s and Z is s by m, then AXA = A implies

(5.1) $CYZB = I_r$,

and XAX = X implies

(5.2) $ZBCY = I_s$.

Consequently, with r = s, ZB = $(CY)^{-1}$ so that

$$AX = BCYZ = B(ZB)^{-1}Z = (BZ)^{\#}BZ$$

and

$$XA = YZBC = Y(CY)^{-1}C = (YC)^{\#}YC,$$

by Lemma 10.

Conversely, since Z and C have full row rank, $(BZ)^{\#}BZB = B$ and $(YC)^{\#}YCY = Y$. Hence AX = $(BZ)^{\#}BZ$ gives AXA = A, and XA = $(YC)^{\#}YC$ gives XAX = X. ∎

It should be noted that the relation in (5.1) is both necessary and sufficient to have AXA = A, and does not require that YZ is a full rank factorization. Dually, (5.2) is both necessary and sufficient to have XAX = X, and BC need not be a full rank factorization. Observe, moreover, that given any matrix, A, with full rank factorization A = BC, then for any choice of Y such that CY has full column rank, taking $Z = (CY)_L B_L$ with $(CY)_L$ any left inverse of CY and B_L any left inverse of B gives a matrix X = YZ such that (5.2) holds. Therefore, we can always construct matrices, X, of any given rank not exceeding the rank of A with XAX = X. On the other hand, given full rank factorizations A = BC and X = YZ such that AXA = A and XAX = X, then for any matrix U with full column rank satisfying CU = 0 and for any matrix V with UVA defined we have

(5.3) $A(X+UV)A = A$.

Now

(5.4) $X + UV = [Y,U] \begin{bmatrix} Z \\ V \end{bmatrix}$

where the first matrix on the right-hand side has full column rank (Exercise 5.6). Thus, for any choice of V such that the second matrix on the right-hand side of (5.4) has full row rank, (5.3) holds and rank (X+UV) > rank (A).

The following example illustrates the construction of matrices, X, of prescribed rank such that A and X satisfy at least one of the conditions AXA = X and XAX = X.

Example 5.1

Let A be the matrix

$$A = \begin{bmatrix} 6 & 4 & 0 \\ 3 & 5 & -3 \\ 3 & 3 & -1 \end{bmatrix}$$

from Example 4.1 with full rank factorization

$$A = BC = \begin{bmatrix} 1 & 2 \\ 2 & 1 \\ 1 & 1 \end{bmatrix} \begin{bmatrix} 0 & 2 & -2 \\ 3 & 1 & 1 \end{bmatrix}.$$

Then rank (A) = 2, and $X_0 = 0$ satisfies $X_0 A X_0 = X_0$ trivially. To construct a matrix, X_1, of rank one such that $X_1 A X_1 = X_1$, note first that

$$(5.5) \qquad B_L = \frac{1}{3}\begin{bmatrix} -1 & 2 & 0 \\ 2 & -1 & 0 \end{bmatrix}$$

is a left inverse of B. Now if $y^H = [3 \quad 4 \quad 5]$, then

$$Cy = \begin{bmatrix} -2 \\ 18 \end{bmatrix}, \quad (Cy)^+ = \frac{1}{164}[-1 \quad 9], \quad z^H = \frac{1}{492}[19 \quad -11 \quad 0]$$

so that

$$X_1 = yz^H = \frac{1}{492}\begin{bmatrix} 57 & -33 & 0 \\ 76 & -44 & 0 \\ 95 & -55 & 0 \end{bmatrix}.$$

To next construct a matrix X_2 of rank two such that $X_2AX_2 = X_2$ (and thus $AX_2A = A$), let

$$Y = \begin{bmatrix} 1 & -1 \\ 1 & 1 \\ 1 & -1 \end{bmatrix}.$$

Then

$$CY = \begin{bmatrix} 0 & 4 \\ 5 & -3 \end{bmatrix}, \quad (CY)^{-1} = \frac{1}{20}\begin{bmatrix} 3 & 4 \\ 5 & 0 \end{bmatrix},$$

and with B_L the left inverse of B in (5.5),

$$Z = (CY)^{-1}B_L = \frac{1}{60}\begin{bmatrix} 5 & 2 & 0 \\ -5 & 10 & 0 \end{bmatrix}$$

so that

$$X_2 = YZ = \frac{1}{30}\begin{bmatrix} 5 & -4 & 0 \\ 0 & 6 & 0 \\ 5 & -4 & 0 \end{bmatrix}.$$

Finally, to construct a matrix, X_3, of rank three such that $AX_3A = A$, let

$$u = \begin{bmatrix} 2 \\ -3 \\ -3 \end{bmatrix}, \quad v = \frac{1}{30}\begin{bmatrix} 0 \\ 0 \\ 1 \end{bmatrix}$$

where $u \varepsilon N(C)$. Then

$$X_3 = X_2 + uv^H = \frac{1}{30}\begin{bmatrix} 5 & -4 & 2 \\ 0 & 6 & -3 \\ 5 & -4 & -3 \end{bmatrix}$$

with det $X_3 = -5$.

That the procedure in Example 5.1 can be extended to construct matrices X of given rank satisfying $(AX)^H = AX$ and at least one of the conditions $AXA = A$ and $XAX = X$ is apparent by observing that CY with full column rank implies BCY has full column rank. Hence, taking $Z = (BCY)^+$, (5.2) holds and $AX = BCY(BCY)^+$ is Hermitian. In the following example we indicate matrices Z in Example 5.1 so that the resulting matrices X_i satisfy $(AX_i)^H = AX_i$, $i = 1,2,3$.

Example 5.2

Given the matrix A and full rank factorization $A = BC$ in Example 5.1, again let $y^H = [3\ \ 4\ \ 5]$. Then

$$Ay = BCy = \begin{bmatrix} 34 \\ 14 \\ 16 \end{bmatrix}, \quad z^H = (Ay)^+ = \frac{1}{804}[17\ \ 7\ \ 8]$$

and

$$X_1 = yz^H = \frac{1}{804}\begin{bmatrix} 51 & 21 & 24 \\ 68 & 28 & 32 \\ 85 & 35 & 40 \end{bmatrix}$$

satisfies $X_1 A X_1 = X_1$ with AX_1 Hermitian and rank $(X_1) = 1$.

Continuing, if we again use

$$Y = \begin{bmatrix} 1 & -1 \\ 1 & 1 \\ 1 & -1 \end{bmatrix}$$

then

$$AY = BCY = \begin{bmatrix} 10 & -2 \\ 5 & 5 \\ 5 & 1 \end{bmatrix}, \quad Z = (AY)^+ = \frac{1}{220}\begin{bmatrix} 16 & 5 & 7 \\ -20 & 35 & 5 \end{bmatrix}$$

and

$$X_2 = YZ = \frac{1}{110}\begin{bmatrix} 18 & -15 & 1 \\ -2 & 20 & 6 \\ 18 & -15 & 1 \end{bmatrix}$$

with $X_2 A X_2 = X_2$, $A X_2$ Hermitian and rank $(X_2) = 2$.

Now taking $u^H = [2 \quad -3 \quad -3]$ as in the previous example and $v^H = 1/110\,[0 \quad 1 \quad 1]$ gives

$$X_3 = X_2 + uv^H = \frac{1}{110}\begin{bmatrix} 18 & -13 & 3 \\ -2 & 17 & 3 \\ 18 & -18 & -2 \end{bmatrix}$$

with $A X_3 A = A$, $A X_3$ Hermitian and $\det X_3 = -10$.

Given any full rank factorization $A = BC$, first choosing a matrix Z so that ZB (and thus ZBC) has full row rank provides a completely dual procedure to that in Example 5.1 in which $Y = C_R(ZB)_R$ with C_R any right inverse of C and $(ZB)_R$ any right inverse of ZB. Taking $Y = (ZBC)^+$ then gives matrices analogous to those in Example 5.2 in which we now have $(X_i A)^H = X_i A$, $i = 1,2,3$.

We conclude this brief introduction to generalized inverses that are not unique by observing that the question of representing all solutions of particular subsets of equations such as $AXA = A$ or $XAX = X$ and AX or XA Hermitian has not been considered. Also, although obvious properties of matrices A and X satisfying $AXA = A$ with AX and XA Hermitian are included in the exercises, the more difficult question when $AXA = A$ is replaced by the nonlinear relation $XAX = X$ is only treated superficially. The interested reader is urged to consult [2] for a detailed discussion of these topics.

Exercises

5.1   Show that any two of the conditions $AXA = A$, $XAX = X$, rank $(X) = $ rank (A) imply the third.

5.2   Show that $XAX = A^+$ if $AXA = A$, $(AX)^H = AX$ and $(XA)^H = XA$.

5.3 Let $A = BC$ and $X = YZ$ where Y and Z^H have full column rank.

  a.   Show that $XAX = X$ and $(AX)^H = AX$ if and only if $BCY = Z^+$. Dually, show that $XAX = X$ and $(XA)^H = XA$ if and only if $ZBC = Y^+$.

  b.   Given the matrix

$$A = \begin{bmatrix} 2 & 1 & -1 \\ 0 & 4 & 3 \\ 1 & -2 & 1 \end{bmatrix},$$

construct a matrix X_1 of rank one such that $X_1 A X_1 = X_1$ and $(AX_1)^H = AX_1$. Also, construct a matrix X_2 of rank two such that $X_2 A X_2 = X_2$ and $(X_2 A)^H = X_2 A$.

5.4 Let $A = BC$ and $X = YZ$ where A is square and B and C^H have full column rank.

  a.   Show that if $AXA = A$ and $XA = AX$, then $B = YZBCB$ and $C = CBCYZ$ where $(CB)^{-1}$ exists.

  b.   Why is it not of interest to consider also the special cases when $(ZB)^{-1}$ or $(CY)^{-1}$ exist?

  c.   What equations must Y and Z satisfy if $XAX = X$ and $AX = XA$?

5.5 Verify that the inverses constructed in Examples 5.1 and 5.2 satisfy the required properties.

5.6 Prove that if $W = [Y,U]$ is any matrix with CY nonsingular and columns of U in $N(C)$ linearly independent, then W has full column rank.

5.7 Prove that if $A = BC$ is any full rank factorization of a square matrix, then $CB = I$ if and only if A is idempotent.

5.8 Show that if $A = BC$ is any full rank factorization and Y is any matrix such that CY is nonsingular, $A^+ = \left[(AY)^+ A\right]^+ (AY)^+$.

Appendix 1:

Hints for Certain Exercises

Chapter 1

Ex. 1.1b: $xy^H = yx^H$ implies $y = \alpha x$ where

$$\alpha = \frac{(x,y)}{||x||^2} \neq 0.$$

Then $\bar{\alpha}xx^H = \alpha xx^H$.

Ex. 1.2c: If $BA^HA = CA^HA$ then $(BA^H - CA^H)A(B^H - C^H) = 0$.

Ex. 1.3a: If P is the matrix with columns $x_1, \ldots, x_n$, and Λ is the diagonal matrix with diagonal elements λ_i, $i = 1, \ldots, n$, $AP = P\Lambda$. Hence $A = (P\Lambda)P^H$ since P is unitary. 1.3c: If A is Hermitian, $\lambda_i E_i = \bar{\lambda}_i E_i$, $i = 1, \ldots, n$.

Ex. 1.5b: If u_k is the column vector with k elements each equal to unity,

$$A_n^{-1} = \begin{bmatrix} -u_{n-1}^H & n \\ I & -u_{n-1} \end{bmatrix}$$

for all $n \geq 2$. 1.5c: Subtract the last row of A_n from each of the preceding rows and expand the determinant using cofactors of the first column. <u>1.5d</u>: A_n^{-1} has all integral elements.

<u>Ex. 1.6a</u>: Let $X = I + \partial xx^H$ and determine ∂ so that $AX = I$. <u>1.6b</u>: $A = 6\left[I + \frac{20}{3}xx^H\right]$ where $x = \frac{1}{\sqrt{40}}u_{40}$. <u>1.6e</u>: $Ax = (1+k)x$ and $Ay = y$.

<u>1.6f</u>: For any $n \geq 2$ the vectors

$$\begin{bmatrix} 1 \\ 1 \\ \cdot \\ \cdot \\ \cdot \\ 1 \end{bmatrix}, \begin{bmatrix} 1 \\ -1 \\ 0 \\ \cdot \\ \cdot \\ 0 \end{bmatrix}, \begin{bmatrix} 1 \\ 1 \\ -2 \\ 0 \\ \cdot \\ 0 \end{bmatrix}, \ldots, \begin{bmatrix} 1 \\ \cdot \\ \cdot \\ \cdot \\ 1 \\ -(n-1) \end{bmatrix}$$

are orthogonal.

<u>Ex. 1.7b</u>: Form XA first.

Chapter 2

<u>Ex. 2.2</u>: $AZ = 0$, and $Z\alpha = 0$ implies $\alpha = 0$.

<u>Ex. 2.5</u>: x_1 is orthogonal to every vector $z \in N(A)$. Hence

$$||x||^2 = ||x_1||^2 + |\alpha_1|^2 + |\alpha_2|^2.$$

<u>Ex. 2.6</u>: Let A be m by n with rank r, so that $\dim N(A) = n-r$. Now assume rank $(A^H A) = k < r$, and let $z_1,\ldots,z_{n-k}$ denote any basis of $N(A^H A)$. Then

$$0 = (z_i, A^H A z_i) = (A z_i, A z_i) = ||A z_i||^2$$

implies $A z_i = 0$, $i = 1,\ldots,n-k$. Hence $\dim N(A) \geq n-k > n-r$, a contradiction.

<u>Ex. 2.10</u>: Use Exercise 2.9 and apply Exercise 1.6a.

<u>Ex. 2.11a</u>: Use Exercise 2.10 to obtain $A^+ b$ and Exercise 2.2 to form $z \in N(A)$. <u>2.11d</u>: In this case

$$A = \begin{bmatrix} I & u_n \\ u_n^H & 0 \end{bmatrix}.$$

<u>Ex. 2.13</u>: $A = uv^H$ is a full rank factorization. Use Theorem 4 and the remarks in the final paragraph of Section 2.2.

<u>Ex. 2.14</u>:
$$\bar{x} = A^+ b + \sum_{i=1}^{n-r} \alpha_i z_i$$

is an orthogonal decomposition of any vector $\bar{x}$. Now take the inner product of $\bar{x}$ with any vector z_i.

<u>Ex. 2.15</u>: For any $i = 1,\ldots,m$, column i of A^+ is the minimal norm solution of $Ax_i = e_i$.

<u>Ex. 2.20e</u>: Use Exercise 1.2c and its dual that $BAA^H = CAA^H$ if and only if $BA = CA$.

Chapter 3

<u>Ex. 3.2b</u>: $\tilde{A} = \begin{bmatrix} W \\ Y \end{bmatrix} \begin{bmatrix} I & W^{-1}X \end{bmatrix}$ is a full rank factorization.

<u>Ex. 3.7a</u>: $d_k^H a_k = 1$ if and only if $c_k \neq 0$.

<u>Ex. 3.11b</u>:
$$A^+ = \frac{1}{q(p+1)} \begin{bmatrix} u_p^H \\ (p+1)I_p - u_p u_p^H \end{bmatrix} \chi u_q^H .$$

Now if $\hat{x} = A^+ y$ is written in terms of components as $\hat{x}^H = (\hat{m}, \hat{t}_1, \ldots, \hat{t}_p)$ then

$$\hat{m} = \frac{1}{q(p+1)} \sum_{i=1}^{p} \sum_{j=1}^{q} y_{ij}$$

and

$$\hat{t}_i = \frac{1}{q} \sum_{j=1}^{q} y_{ij} - \frac{1}{q(p+1)} \sum_{i=1}^{p} \sum_{j=1}^{q} y_{ij}, \quad i = 1,\ldots,p.$$

<u>3.11c</u>: With dim $N(A) = 1$ and $z = \begin{bmatrix} -1 \\ u_p \end{bmatrix} \epsilon N(A)$, all solutions of $Ax = y$ can be written in terms of components as $m = \hat{m} - \alpha$ and $t_i = \hat{t}_i + \alpha$, $i = 1,\ldots,p$, where α is arbitrary.

If $\sum_{i=1}^{p} t_i = 0$, then

$$\alpha = \frac{-1}{p} \sum_{i=1}^{p} \hat{t}_i = \frac{-1}{pq} \sum_{i=1}^{p} \sum_{y=1}^{q} y_{ij} + \frac{1}{q(p+1)} \sum_{i=1}^{p} \sum_{y=1}^{q} y_{ij}$$

$$= - \frac{1}{pq(p+1)} \sum_{i=1}^{p} \sum_{j=1}^{q} y_{ij}$$

so that

$$\hat{\hat{m}} = \frac{1}{pq} \sum_{i=1}^{p} \sum_{j=1}^{q} y_{ij}$$

and

$$\hat{\hat{t}}_i = \frac{1}{q} \sum_{j=1}^{q} y_{ij} - \frac{1}{pq} \sum_{i=1}^{p} \sum_{j=1}^{q} y_{ij}.$$

Ex. 3.13: $\lambda x \bigtimes \mu y = \lambda \mu (x \bigtimes y)$.

Ex. 3.14a: $I_{np} = I_n \bigtimes I_p$ and $\frac{1}{p} u_p^H = u_p^+$.

3.14b: Let $z_1, \ldots, z_n$ be any complete orthonormal set of eigenvectors of $I_n - W^+W$ where $z_1, \ldots, z_r$ correspond to eigenvalue $\lambda = 1$ and $z_{r+1}, \ldots, z_n$ correspond to eigenvalue $\lambda = 0$. Combine these vectors with those in the hint for Exercise 1.6f.

Ex. 3.15: Use Gauss elimination to reduce T to block form

$$\begin{bmatrix} I_n \bigtimes e_p^H \\ 0 \bigtimes u_p^H \\ W \bigtimes [0, I_{p-1}] \end{bmatrix}.$$

Ex. 3.17: $z_i \bigtimes y_j$ corresponds to eigenvalue one if and only if

$$T(z_i \bigtimes y_j) = 0.$$

Ex. 3.20: $AX = C$ and $XB = D$ consistent implies $AA^+C = C$ and $DB^+B = D$.

Chapter 4

Ex. 4.2a: $(B^H B)^{-1} B^H = (C^H B)^+ C^H$. 4.2b: The relation is reflexive, by Lemma 5(e). If C is alias to B, then $C = BB^+ C = B^{+H}(B^H C)$ is a full rank factorization and the relation is symmetric since $(B^{+H})^+ = B^H$. Transitivity follows by a similar type of argument.

Ex. 4.9: With $W = B^H$, $(B_W)_d = B(B^H B)_d^2 = B[(B^H B)^+]^2$.

Ex. 4.11: Use an argument similar to that one employed to establish uniqueness in (2.2).

Ex. 4.16: If $A = BC$ is a full rank factorization and A has index one,

$$A_L = C^+(CB)^{-1}C \quad \text{and} \quad A_R = B(CB)^{-1}B^+,$$

by Exercise 4.14. Then

$$A_L A = C^+ C = A^+ A, \quad A_R^n = B(CB)^{-n}B^+ \quad \text{and} \quad A^n A_R^n = BB^+ = AA^+$$

for all $n \geq 1$.

Ex. 4.17: Show first that $X = B(WB)_d^P$ satisfies all six equations. Then show that the first set of three equations implies the second set, and that the second set implies X has the given form.

Ex. 4.18: If $A_d^2 = B_d^2$ then $A_d B = AB_d$, so that $A_d = A_d BB_d$ and $B_d = A_d AB_d$.

Chapter 5

Ex. 5.2: Use both the direct and dual form of Exercise 4.10 with $\ell = 1$.

Ex. 5.3a: Applying Exercise 4.10 to $XAX = X$ and $(AX)^H = AX$ gives $AX = X^+ X = Z^+ Z$.

Ex. 5.4a: B has full column rank. 5.4b: Then $X = A^\#$.

Ex. 5.8: $AY = B(CY)$ is a full rank factorization.

Appendix 2:
Selected References

1. Anderson, R.L. and Bancroft, T.A. 1952. Statistical Theory in Research. New York: McGraw-Hill.

2. Ben-Israel, A. and Greville, T.N.E. 1974. Generalized Inverses: Theory and Applications. New York: Wiley.

3. Drazin, M.P. 1958. Pseudo-inverses in associative rings and semi-groups. Amer. Math. Monthly 65:506-513.

4. Graybill, F.A. 1961. An Introduction to Linear Statistical Models, Volume I. New York: McGraw-Hill.

5. Greville, T.N.E. 1961. Note on fitting of functions of several independent variables. SIAM J. Appl. Math. 9:109-115 (Erratum, p. 317).

6. Hadley, G. 1962. _Linear Programming_. Massachusetts: Addison-Wesley.

7. Halmos, P.R. 1958. _Finite-Dimensional Vector Spaces_. New Jersey: Van-Nostrand.

8. Moore, E.H. 1920. On the reciprocal of the general algebraic matrix (abstract). _Bull. Amer. Math. Soc._ 26:394-395.

9. Noble, B. 1969. _Applied Linear Algebra_. New Jersey: Prentice Hall.

10. Penrose, R. 1955. A generalized inverse for matrices. _Proc. Camb. Phil. Soc._ 51:406-413.

11. Rao, C.R. and Mitra, S.K. 1971. _Generalized Inverses of Matrices and Its Applications_. New York: Wiley.

12. Strang, G. 1976. _Linear Algebra and Its Applications_. New York: Academic Press.

13. Thrall, R.M. and Tornhiem, L. 1957. _Vector Spaces and Matrices_. New York: Wiley.

Index to
Principal Definitions

Index to
Principal Definitions